Infinite Dimensional

# Lie Groups

## in Geometry and
## Representation Theory

# Infinite Dimensional

# Lie Groups

## in Geometry and Representation Theory

Washington, DC, USA
17–21 August 2000

Editors

**Augustia Banyaga**
*Pennsylvania State University, USA*

**Joshua A Leslie**
*Howard University, USA*

**Thierry Robart**
*Howard University, USA*

**World Scientific**
*New Jersey • London • Singapore • Hong Kong*

*Published by*

World Scientific Publishing Co. Pte. Ltd.

5 Toh Tuck Link, Singapore 596224

*USA office:* 27 Warren Street, Suite 401-402, Hackensack, NJ 07601

*UK office:* 57 Shelton Street, Covent Garden, London WC2H 9HE

**British Library Cataloguing-in-Publication Data**
A catalogue record for this book is available from the British Library.

**INFINITE DIMENSIONAL LIE GROUPS IN GEOMETRY AND REPRESENTATION THEORY**

ISBN-13 978-981-238-068-5
ISBN-10 981-238-068-X

# Preface

This volume contains papers delivered at the occasion of the 2000 Howard fest on *Infinite Dimensional Lie Groups in Geometry and Representation Theory*. The five day International Conference was held on the main campus of Howard University from the 17th to the 21st August, 2000. We believe that the collected papers, by presenting important recent developments, should offer a valuable source of inspiration for advanced graduate students and/or established researchers in the field. All papers have been refereed.

## A Short overview:

The book opens with a topological characterization of regular Lie groups in the context of Lipschitz-metrizable groups, a class that contains all strong ILB-Lie groups introduced by Omori in the early seventies (Josef Teichmann). It then treats the integrability problem of various important infinite dimensional Lie algebras: a canonical approach of the general integration problem based on charts of the second kind and illustrated with the isotropy group of local analytic vector fields is described (Thierry Robart); a result of Goodman and Wallach is extented to a very large class of Kac-Moody algebras associated to generalized symmetrizable Cartan matrices (Joshua Leslie); and within the framework of bounded geometry the known Lie group structure of invertible Fourier integral operators on compact manifolds is shown to hold equally for open manifolds (Rudolf Schmid).

The volume contains also important contributions at the forefront of modern geometry. Firstly, the main properties of Leibniz algebroids are studied here (Aissa Wade). The concept of Leibniz algebroid was introduced recently in the study of Nambu-Poisson structures. As weakened version of that of Lie algebroid, it represents a far-reaching generalization of the classical concept of Lie algebra.

There are five papers devoted to locally conformal symplectic geometry (Augustin Banyaga, Stefan Haller), contact geometry (Philippe Rukimbira), smooth orbifold structures (Joseph E. Borzellino and Victor Brunsden) and the equivalence problem of Poisson and symplectic structures (Augustin Banyaga and Paul Donato). There the focus is mainly on the interaction between the studied structures and their associated infinite dimensional Lie groups of symmetries in the spirit of the 1872 Erlanger Programme of Felix Klein. It is shown in particular that the automorphism groups of locally conformal symplectic structures and of smooth orbifolds determine the corresponding structures. These strong results have many applications. They can

v

be used among other for classification purpose; for instance locally conformal symplectic structures are classified according to a certain homomorphism (the Lee homomorphism) on their automorphism groups. Unit Reeb fields on contact manifolds, viewed as maps from the manifold into its unit tangent bundle, are characterized as harmonic maps or minimal embeddings under certain conditions.

The book concludes with penetrating remarks concerning the concept of amenability, infinite dimensional groups and representation theory (Vladimir Pestov).

## List of Participants/Authors

Augustin Banyaga (Penn State University), Joe Borzellino (California State Polytechnic University), Michel Boyom (Université de Montpellier II, France), Victor Brunsden (Penn State University), Paul Donato (Centre de Mathématiques et d'Informatique, Marseille, France), Stefan Haller (University of Vienna, Austria), Patrick Iglesias (Centre de Mathématiques et d'Informatique, Marseille, France), Joshua Leslie (Howard University), Peter Michor (Vienna University, Austria), Hideki Omori (Science University of Tokyo, Japan), Vladimir Pestov (Victoria University of Wellington, New Zealand), Tudor Ratiu (Ecole Polytechnique Federale de Lausanne, Switzerland), Thierry Robart (Howard University), Philippe Rukimbira (Florida International University), Rudolf Schmid (Emory University), Josef Teichmann (Technische Universität Wien, Austria), Aissa Wade (Penn State University).

## Contributions

- *Inheritance properties for Lipschitz-metrizable Frölicher groups* by Josef Teichmann,

- *Around the exponential mapping* by Thierry Robart,

- *On a solution to a global inverse problem with respect to certain generalized symmetrizable Kac-Moody Lie algebras* by Joshua Leslie,

- *The Lie group of Fourier integral operators on open manifolds* by Rudolf Schmid,

- *On Some properties of Leibniz Algebroids* by Aissa Wade,

- *On the geometry of locally conformal symplectic manifolds* by Augustin Banyaga,

- *Some properties of locally conformal symplectic manifolds* by Stefan Haller,

- *Criticality of unit contact vector fields* by Philippe Rukimbira,

- *Orbifold Homeomorphism and Diffeomorphism Groups* by Joseph E. Borzellino and Victor Brunsden,

- *A note on Isotopies of Symplectic and Poisson Structures* by Augustin Banyaga and Paul Donato,

- *Remarks on actions on compacta by some infinite-dimensional groups* by Vladimir Pestov,

## Acknowledgment

We would like to express our deepest gratitude to the National Security Agency. This Conference wouldn't have been possible without its generous support.

We also wish to thank all the participants, authors, referees and colleagues for their various and irreplaceable contributions, specially Aissa Wade and Philippe Rukimbira for their constant help during the preparation of this volume.

A. Banyaga, J. Leslie & T. Robart
May 2, 2002

# Contents

# INHERITANCE PROPERTIES FOR
# LIPSCHITZ-METRIZABLE FRÖLICHER GROUPS

JOSEF TEICHMANN

*Institute of financial and actuarial mathematics, Technical University of Vienna,*
*Wiedner Hauptstraße 8-10, A-1040 Vienna, Austria*
*E-mail: josef.teichmann@fam.tuwien.ac.at*

Frölicher groups, where the notion of smooth map makes sense, are introduced. On Frölicher groups we can formulate the concept of Lipschitz metrics. The resulting setting of Frölicher-Lie groups can be compared to generalized Lie groups in the sense of Hideki Omori. Furthermore Lipschitz-metrics on Frölicher groups allow to prove convergence of approximation schemes for differential equations on Lie groups. We prove several inheritance properties for Lipschitz metrics.

## 1 Introduction

Lipschitz-metrizable groups have been introduced in [6] to show that regularity of Lie groups (see [1] for all necessary details on Lie groups) is closely connected to some approximation procedures possible on Lie groups. The convergence of these approximation schemes is guaranteed by Lipschitz metrics.

In the work of Hideki Omori et al. the beautiful framework of strong $ILB$-Lie groups is provided (see [2] for example), where the problem of regularity is solved by analytic assumption on the group-multiplication in the charts. The advantage of Lipschitz-metrizable groups is that the notion is "inner", i.e. formulated on the Lie group itself without charts. All strong $ILB$-groups are Lipschitz-metrizable regular groups (see [6], Corollary 2.10). Given a Lipschitz metrizable Lie group we can – by approximation schemes – characterize the existence of exponential and evolution maps. This can also be applied to solve more general equations as stochastic differential equations on Lipschitz-metrizable Lie groups.

In this note we motivate the method of Lipschitz metrics and investigate roughly the inheritance properties of Lipschitz-metrizable Frölicher-Lie groups.

**Definition 1.** *A non-empty set $X$, a set of curves $C_X \subset Map(\mathbb{R}, X)$ and a set of mappings $F_X \subset Map(X, \mathbb{R})$ are called a Frölicher space if the following conditions are satisfied:*

*1. A map $f : X \to \mathbb{R}$ belongs to $F_X$ if and only if $f \circ c \in C^\infty(\mathbb{R}, \mathbb{R})$ for $c \in C_X$.*

2. *A curve $c : X \to \mathbb{R}$ belongs to $C_X$ if and only if $f \circ c \in C^\infty(\mathbb{R}, \mathbb{R})$ for $f \in F_X$*

Let $X$ be a Frölicher space, then $C_X$ is called the set of smooth curves, $F_X$ the set of smooth real valued functions. Maps between Frölicher spaces are called smooth if their composition with any smooth curve is smooth. Let $X, Y$ be Frölicher spaces then the set of smooth maps $C^\infty(X, Y)$ has a natural structure of a Frölicher space due to the following requirement:

$$C^\infty(X, Y) \overset{C(f,c)}{\to} C^\infty(\mathbb{R}, \mathbb{R}) \overset{\lambda}{\to} \mathbb{R}$$

is a smooth map for $c \in C_X$, $f \in F_Y$ and $\lambda \in C^\infty(\mathbb{R}, \mathbb{R})'$, where $C(f, c)(\phi) := f \circ \phi \circ c$. The Frölicher space structure generated by these smooth real valued maps is the canonical structure on $C^\infty(X, Y)$. The topology $c^\infty X$ is given by the initial topology with respect to smooth curves (for all necessary details see [1], chapter V).

A Frölicher group is an abstract group $G$ with the structure of a Frölicher space, such that multiplication and inversion are smooth and $c^\infty G$ is a topological group. Furthermore we assume that the smooth functions separate points (smoothly Hausdorff, see [1], chapter VI).

## 2 Lipschitz-metrizable Frölicher groups

In the sequel of the article we shall need the following approximation theorem for product integrals. They exist if their approximations lie uniformly on compact sets in bounded sets, for the notions of convenient calculus we refer to [1]. For the approximation theorem and the notion of product integrals see [2], [4] and [6].

**Definition 2 (Product integral).** *Let $A$ be a convenient algebra. Given a smooth curve $X: \mathbb{R} \to A$ and a smooth mapping $h : \mathbb{R}^2 \to A$ with $h(s, 0) = e$ and $\frac{\partial}{\partial t} h(s, 0) = X(s)$, then we define the following finite products of smooth curves*

$$p_n(a, t, h) := \prod_{i=0}^{n-1} h\left(a + \frac{(n-i)(t-a)}{n}, \frac{t-a}{n}\right)$$

*for $a, t \in \mathbb{R}$. If $p_n$ converges in all derivatives to a smooth curve $c : \mathbb{R} \to A$, then $c$ is called the product integral of $X$ or $h$ and we write $c(a, t) = \prod_a^t \exp(X(s)ds)$ or $c(a, t) =: \prod_a^t h(s, ds)$. The case $h(s, t) = c(t)$ with $p_n(0, t, h) = c(\frac{t}{n})^n$ is referred to as simple product integral.*

**Theorem 1 (Approximation theorem).** *Let $A$ be convenient algebra. Given a smooth curve $X : \mathbb{R} \to A$ and a smooth mapping $h : \mathbb{R}^3 \to A$ with $h(u,r,0) = e$ and $\frac{\partial}{\partial t}h(u,r,0) = X_u(r)$. Suppose that for every fixed $s_0 \in \mathbb{R}$, there is $t_0 > s_0$ such that $p_n(u,s,t,h)$ is bounded in $A$ on compact $(u,s,t)$-sets and for all $n \geq 1$. Then the product integral $\prod_s^t h(u,r,dr)$ exists and the convergence is Mackey in all derivatives on compact $(u,s,t)$-sets. Furthermore the product integral is the right evolution of $X_u$, i.e.*

$$\frac{\partial}{\partial t} \prod_s^t h(u,r,dr) = X_u(t)\frac{\partial}{\partial t} \prod_s^t h(u,r,dr),$$

$$\prod_s^s h(u,r,dr) = e.$$

**Remark 1.** *The hypothesis on the product integrals will be referred to as* **boundedness condition***. For the proof see* [4]*, Theorem 2.2.*

The definition of product integrals on Frölicher groups $G$ is done in the same spirit as on convenient algebras, see [6] for details. It yields the natural non-commutative extension of integration.

**Definition 3.** *Let $G$ be a Frölicher group. Given a smooth mapping $h : \mathbb{R}^2 \to G$ with $h(s,0) = e$, then we define the following finite products of smooth curves*

$$p_n(s,t,h) := \prod_{i=0}^{n-1} h(s + \frac{(n-i)(t-s)}{n}, \frac{t-s}{n})$$

*for $s,t \in \mathbb{R}$. If $p_n$ converges in the smooth topology of $G$ uniformly on compact sets to a continuous curve $c : \mathbb{R} \to G$, then $c$ is called the* product integral *of $h$ and we write $c(s,t) =: \prod_s^t h(u,du)$. If $h(s,t) = c(t)$, then the product integral $p_n(0,t,h) = c(\frac{t}{n})^n$ is called* simple product integral.

**Remark 2.** *Here we need the assumption that $c^\infty G$ is a topological group, since we want to talk about uniform convergence and completeness of the uniform space $c^\infty G$.*

The right regular representation $\rho$ of a Frölicher group $G$

$$\rho : G \to L(C^\infty(G,\mathbb{R}))$$

$$g \mapsto (f \mapsto f(.g))$$

in the bounded operators on $C^\infty(G,\mathbb{R})$ is initial and smooth (see [1] for the notion of initial maps). We shall apply this "linearization" in the sense of Lemma 1 several times in the article (see [6] for further details). This is the main link from the existence theorem for evolutions via Trotter approximations to the general case on Frölicher groups.

4

**Lemma 1.** *Let $G$ be a Frölicher group, then each product $p_n(s,t,h)$ and the limit - if it exists - $\prod_s^t h(u,du)$ is smooth. The propagation condition*

$$\prod_t^r h(u,du) \prod_s^t h(u,du) = \prod_s^r h(u,du)$$

*is satisfied for all $r,s,t$.*

*Proof.* By the left regular representation $\rho$ on $G$ we get that the product integral

$$\lim_{n\to\infty} p_n(s,t,\rho\circ h)$$

exists in $C^\infty(\mathbb{R}^2, L(C^\infty(G,\mathbb{R})))$, since that image of a sequentially compact set under a smooth mapping is in particular bounded in the convenient algebra $L(C^\infty(G,\mathbb{R}))$. The set formed by $p_n(s,t,h)$ and $\prod_s^t h(u,du)$ on compact $(s,t)$-sets is sequentially compact due to uniform convergence. Consequently we are in the hypotheses of the approximation theorem Theorem 1, which allows the conclusion of smoothness of $\prod_s^t h(u,du)$, since

$$\rho\left(\prod_s^t h(u,du)\right)(f) = \lim_{n\to\infty} p_n(s,t,\rho\circ h)(f)$$

is smooth and we can evaluate the both sides at $e$ for any $f \in C^\infty(G,\mathbb{R})$. Therefore – by the definition of Frölicher spaces – we get the desired conclusion for

$$(s,t) \mapsto f\left(\prod_s^t h(u,du)\right)$$

is smooth. The propagation condition finally follows from the definition of product integrals and the continuity of multiplication. $\square$

Lipschitz-metrizable Frölicher groups are the adequate framework for the application of Trotter approximations.

**Definition 4 (Lipschitz-metrizable groups).** *Let $G$ be a Frölicher group, such that $c^\infty G$ is a topological group. $G$ is called Lipschitz-metrizable if there is a family of right invariant halfmetrics $\{d_\alpha\}_{\alpha\in\Omega}$ on $G$ with the following properties:*

1. *For all sequences $\{x_n\}_{n\in\mathbb{N}}$:*

$$\forall \alpha \in \Omega : d_\alpha(x_k,x_l) \to 0 \iff \{x_n\}_{n\in\mathbb{N}} \text{ is converging in } G.$$

2. *For all smooth mappings* $c : \mathbb{R}^2 \to G$ *with* $c(s,0) = e$, *there is on each compact* $(s,t)$-*set a constant* $M_\alpha$ *such that*

$$d_\alpha(c(s,t),e) < M_\alpha t.$$

On Lipschitz-metrizable groups we can formulate a condition for the existence of evolution maps. Therefore we need the notion of "touching at a $e$". Given $h_{1,2} : \mathbb{R}^2 \to G$ with $h_{1,2}(s,0) = e$, then we say that $h_1$ touches $h_2$ at $e$ if for each compact $(s,t)$-set there is a constant $M_\alpha$ such that

$$d_\alpha(h_1(s,t), h_2(s,t)) \leq M_\alpha t^2$$

on the given compact set. A "direction" in $G$ is simply given by an equivalence class of curves $c : \mathbb{R} \to G$ with $c(0) = e$ under the equivalence relation of "touching at $e$". We denote the set of equivalence classes by $\mathfrak{g}$. A smooth group $T : \mathbb{R} \to G$ is a *smooth group homomorphism* or *smooth one-parameter subgroup*, a *right evolution* $d : \mathbb{R}^2 \to G$ is a smooth mapping with $d(s,s) = e$, such that $d(s,t)d(r,s) = d(r,t)$ for all $r,s,t \in \mathbb{R}$. Given a smooth curve $c : \mathbb{R} \to G$ with $c(0) = e$, then $(s,t) \mapsto c(t)c(s)^{-1}$ is a smooth right evolution.
**Theorem 2.** *Let $G$ be a Lipschitz-metrizable Frölicher group, $h : \mathbb{R}^2 \to G$ a smooth mapping with $h(s,0) = e$, a smooth curve $c : \mathbb{R} \to G$ with $c(0) = e$ such that $h$ touches the right evolution associated to $c$ at $e$, then the product integral $\prod_0^t h(s,ds)$ exists and equals $c(t)$. The following estimates are valid for the Lipschitz-metrics $d_\alpha$:*

$$d_\alpha(p_i(s_3,t,c)(s_1)p_n(s_2,t+s_2,c)(s_1)c(s_1,s_2,t)^{-1}p_i(s_3,t,c)(s_1)^{-1},e) \leq M_\alpha t^2$$

*for all $i,n \in \mathbb{N}_+$ on compact $(s_1,s_2,s_3,t)$-sets given the smooth mapping $d : \mathbb{R}^3 \to G$ with $c(s_1,s_2,0) = e$.*
**Theorem 3.** *Let $G$ be a Lipschitz-metrizable Frölicher group with $c^\infty G$ a topological group. For all smooth mappings $c : \mathbb{R}^3 \to G$ with $c(s_1,s_2,0) = e$ the following estimate is valid*

$$d_\alpha(p_i(s_3,t,c)(s_1)p_n(s_2,t+s_2,c)(s_1)c(s_1,s_2,t)^{-1}p_i(s_3,t,c)(s_1)^{-1},e) \leq M_\alpha t^2$$

*for all $n \in \mathbb{N}$ on compact $(s_1,s_2,s_3,t)$-sets. Then all product integrals exist and depend smoothly on smooth parameters.*
**Remark 3.** *In the case of convenient Lie groups this condition can be interpreted as condition on the "second derivatives" of the multiplication mapping. Both theorems are proved in* [6]. *These Theorems could also be applied to obtain "regular" solutions of Stratonovich differential equations on compact (or noncompact) manifolds by product integration of stochastic vector fields. This will be worked out elsewhere.*

Given a Frölicher group $G$, then we define its tangent space $T_eG$ in the following way: $\rho : G \to L(C^\infty(G,\mathbb{R}))$ is smooth, so

$$T_eG := \{\frac{d}{dt}|_{t=0}\rho(c(t)) \text{ for } c : \mathbb{R} \to G \text{ smooth with } c(0) = e\}.$$

This is a linear subspace of $L(C^\infty(G,\mathbb{R}))$. Given $\Phi : G \to H$ smooth, then the tangent map $\Phi' : T_eG \to T_eH$ can be defined in the following way: Given $X \in T_eG$, then there is $c : \mathbb{R} \to G$ with $c(0) = e$ such that $X = \frac{d}{dt}|_{t=0}\rho(c(t))$, we define $(\Phi'(X)\cdot f)(h) := \frac{d}{dt}|_{t=0}f(\Phi(hc(t)))$ for all $h \in H$ and $f \in C^\infty(H,\mathbb{R})$. This does not depend on the representative since $f(h\Phi(.))$ for fixed $h \in H$ is a smooth function on $G$ and two representatives $c_1$ and $c_2$ with $c_1(0) = c_2(0) = e$ are equal if for all smooth functions $f : G \to \mathbb{R}$ the equality $\frac{d}{dt}|_{t=0}f(c_1(t)) = \frac{d}{dt}|_{t=0}f(c_2(t))$ holds. The tangent map of the conjugation $conj(g)(h) := ghg^{-1}$ for $g, h \in G$ is denoted by $Ad(g)$ and defines a group isomorphism $Ad : G \to Lin(T_eG)$. Notice that these definitions do not require charts, however, we are interested in more analytic structures on $G$ itself to be able to apply the above linearizations adequately.

**Definition 5.** *A Lipschitz-metrizable Frölicher group $G$ is called Frölicher-Lie group if $\mathfrak{g}$ carries a convenient structure by defining*

$$[c_1] + [c_2] = [c_1 \cdot c_2],$$
$$\lambda \cdot [c] = [c(\lambda.)]$$

*with the norm-functions on $\mathfrak{g}$, namely*

$$p_\alpha([c]) := \underline{\lim}_{t\downarrow 0}\frac{d_\alpha(c(t),e)}{t}.$$

*Furthermore the mapping $s \mapsto [t \mapsto h(s,t)]$ is assumed to be smooth for any smooth mapping $h : \mathbb{R} \to G$ with $h(s,0) = e$ and this are all smooth mappings. Finally there is a diffeomorphism*

$$Evol^r : C^\infty(\mathbb{R},\mathfrak{g}) \to C^\infty(\mathbb{R},\{0\};G,\{e\})$$

*defined by $Evol^r(s \mapsto [h(s,.)]) = \prod_0^t h(s,ds)$, where $h : \mathbb{R}^2 \to G$ a smooth mapping with $h(s,0) = e$. The inverse is given by $\delta^r c$, the right logarithmic derivative. Furthermore we denote $Evol^r(t \mapsto tX) =: \exp(tX)$, the exponential map.*

**Remark 4.** *On strong ILB-groups we are given the structure of a Lipschitz-metrizable Frölicher-Lie group, see [6], Corollary 2.10. Due to these properties one can view Lipschitz-metrics on Lie groups as non-abelian generalizations of seminorms on convenient locally convex spaces (which are abelian Lie groups).*

**Lemma 2.** *Let $G$ be a Frölicher-Lie group such that $\mathfrak{g}$ carries the above convenient structure, then $\rho : G \to L(C^\infty(G, \mathbb{R}))$ induces a unique smooth mapping $\rho' : \mathfrak{g} \to L(C^\infty(G, \mathbb{R}))$ with $(\rho'([c]) \cdot f)(g) := \frac{d}{dt}|_{t=0} f(gc(t))$.*

*Proof.* Take a curve $c : \mathbb{R} \to G$ with $c(0) = e$, then $\lim_{n \to \infty} c(\frac{t}{n})^n =: T_t$ defines a smooth group. $\rho(c(\frac{t}{n})^n) = [\rho(c(\frac{t}{n}))]^n$ converges to $\rho(T_t)$ in the convenient topology of $L(C^\infty(G, \mathbb{R}))$, so in particular $f(gc(\frac{t}{n})^n) \to f(gT_t)$ for every $g \in G$ on compact $t$-intervals in any derivative, so

$$\frac{d}{dt}|_{t=0} f(gc(t)) = \frac{d}{dt}|_{t=0} f(gc(\frac{t}{n})^n) = \frac{d}{dt}|_{t=0} f(gT_t).$$

So $\rho'$ is well defined on every equivalence class and if we take a smooth curve $X$ into $\mathfrak{g}$ we see by the exponential map that

$$((\rho' \circ X)(s) \cdot f)(g) = \frac{d}{dt}|_{t=0} f(g \exp(tX(s))),$$

which is smooth. $\qquad\square$

Given two Frölicher-Lie groups $G$ and $H$ and a smooth homomorphism $\Phi : G \to H$, then $\phi := \Phi'$ defined via $\Phi'([c]) = [\Phi \circ c]$ is a smooth linear mapping. Since by regularity every equivalence class $[c]$ contains a smooth group, we see that $\Phi'$ is well defined and smooth. If $G$ is a Frölicher-Lie group, then we can define the Lie bracket on $\mathfrak{g}$: Given $g \in G$, then there the conjugation $conj(g)$ is a smooth mapping. $conj(g)' =: Ad(g)$ is smooth and depends smoothly on $g$. Given $c : \mathbb{R} \to G$ with $c(0) = e$, then $[[c], [d]] := \frac{d}{dt}|_{t=0} Ad(c(t)) \cdot [d]$.

**Lemma 3.** *Let $G$ be a Frölicher-Lie group, then $\delta^r(cd)(s) = \delta^r c(s) + Ad(c(s)) \cdot \delta^r d(s)$ for smooth curves $c$ and $d$.*

*Proof.* Remark that $\delta^r c$ for a smooth curve $c$ is defined by "touching at $e$" of $t \mapsto c(t + s)c(s)^{-1}$. $[Ad(c(s))\delta^r d](s)$ is defined via "touching at $e$" of $t \mapsto conj(c(s))d(t + s)(conj(c(s))d(s))^{-1}$, so we see immediately

$$[t \mapsto c(t + s)d(t + s)d(s)^{-1}c(s)^{-1}]$$
$$= [t \mapsto c(t + s)c(s)^{-1} conj(c(s))d(t + s)(conj(c(s))d(s))^{-1}],$$

which yields the formula. $\qquad\square$

**Remark 5.** *A Lipschitz-metrizable Frölicher group such that the space $\mathfrak{g}$ is a convenient vector space with the above structures, in particular the space $C^\infty(\mathbb{R}, \mathfrak{g})$ coinciding with*

$$\{s \mapsto [t \mapsto h(s, t)] \text{ for } h : \mathbb{R}^2 \to G \text{ a smooth mapping with } h(s, 0) = e\},$$

8

*is a Frölicher-Lie group if and only if the conditions of Theorem 3 are valid. This is proved by Theorem 2 and Theorem 3. In particular this is true for all Lipschitz-metrizable convenient Lie groups. Notice that further elements of Lie theory can be developed similarly to [1].*

## 3 Inheritance properties for Frölicher-Lie groups

A virtual Frölicher-Lie subgroup $H$ of a Frölicher-Lie group $G$ is defined to a subgroup together with $\mathfrak{h}$, a $c^\infty$-closed subalgebra of $\mathfrak{g}$ such that $Evol^r : C^\infty(\mathbb{R}, \mathfrak{h}) \to C^\infty(\mathbb{R}, \{0\}; H, \{e\})$ is a diffeomorphism. A Frölicher-Lie subgroup is a closed, virtual Frölicher-Lie subgroup.

**Theorem 4.** *We obtain the following inheritance properties:*

1. *Let $G_i$ for $i \in I$ be an arbitrary set of Frölicher groups, then $\prod_{i \in I} G_i$ is a Frölicher group.*

2. *Let $G$ be a Frölicher-Lie group and $H$ a closed subgroup, then $H$ is a Frölicher-Lie subgroup.*

3. *Let $G$ be a Frölicher-Lie group and $H$ a normal (closed) Frölicher-Lie subgroup, then $G/H$ is a Frölicher group and the canonical projection $\pi : G \to G/H$ is smooth. If $\mathfrak{g}/\mathfrak{h}$ is convenient with lifting property, i.e. a smooth curve to $\mathfrak{g}/\mathfrak{h}$ has a smooth lift along $\pi'$ and $G/H$ has the lifting property, then $G/H$ is a Frölicher-Lie group with Lie algebra given by $\mathfrak{g}/\mathfrak{h}$.*

4. *Let $G$ be a Frölicher Lie group and $\mathfrak{h}$ a $c^\infty$-closed subalgebra of $\mathfrak{g}$, then there is a unique, smoothly connected, virtual Frölicher-Lie subgroup $H$ with Lie algebra $\mathfrak{h}$.*

**Remark 6.** *One might wonder why we do not treat finite or even infinite products of Frölicher-Lie groups, since we developed the whole theory for the infinite dimensional usage. However, since $c^\infty(G_1 \times G_2)$ is in general finer than $c^\infty G_1 \times c^\infty G_2$, we cannot hope for good compatibility of the product topology and the $c^\infty$-topology.*

*Proof.* Let $G$ be a Frölicher-Lie group and $H$ a smoothly connected, closed subgroup, then we can define the future Lie algebra

$$\mathfrak{h} := \{X \in \mathfrak{g} \text{ such that } \exp(sX) \in H \text{ for all } s \in \mathbb{R}\}$$

as a subset of $\mathfrak{g}$. If $X, Y \in \mathfrak{h}$, then

$$\exp(s(X+Y)) = \lim_{n \to \infty} (\exp(\frac{s}{n}X)\exp(\frac{s}{n}Y))^n$$

lies in $H$ for all times $s$ by closedness and regularity, furthermore $\lambda X \in \mathfrak{h}$ for any real number $\lambda$. Consequently $\mathfrak{h}$ is a linear subspace of $\mathfrak{g}$. By smoothness of the evolution map we obtain its continuity: so $X_n \to X$ in $\mathfrak{g}$ with $X_n \in \mathfrak{h}$ leads to $\exp(sX_n) \to \exp(sX)$ pointwise in the $c^\infty$-topology of $G$, consequently $X \in \mathfrak{h}$, so $\mathfrak{h}$ is $c^\infty$-closed in $\mathfrak{g}$ and hence convenient. The closed subset $H$ carries the natural structure of a Frölicher group, the metrics can be restricted and so we obtain a Lipschitz-metrizable Frölicher group. Since $\mathfrak{h}$ is convenient and the restriction of the Evolution map is smooth, we are ready.

If $H$ is normal, then $\mathfrak{h}$ is an ideal, since $conj(g)H = H$ for all $g \in G$, so $Ad(g)(\mathfrak{h}) \subset \mathfrak{h}$ for all $g \in G$, therefore $[\mathfrak{g}, \mathfrak{h}] \subset \mathfrak{h}$. We denote $\pi : G \to G/H$ and $\pi' : \mathfrak{g} \to \mathfrak{g}/\mathfrak{h}$ the projections. We have to define the structure of a Frölicher space on $G/H$ by defining its smooth functions, $f : G/H \to \mathbb{R}$ is smooth if $f \circ \pi$ is smooth. In particular curves of the type $\pi \circ c$ for $c : \mathbb{R} \to G$ are smooth, in particular $\pi$ is smooth. The multiplication and inversion are therefore smooth. We investigate now the mapping $\pi_* \circ Evol^r : C^\infty(\mathbb{R}, \mathfrak{g}) \to C^\infty(\mathbb{R}, \{0\}; G/H, \{H\})$. Given $X : \mathbb{R} \to \mathfrak{g}$, then $\pi_* \circ Evol^r(X) = \pi(Evol^r(X))$ only depends on $\pi' \circ X$, since any other smooth lift of this curve is a limit of product integrals with exponentials and for exponentials we have

$$\pi \exp(s(X + Y)) = \pi \exp(sX)$$

for $X \in \mathfrak{g}$ and $Y \in \mathfrak{h}$ by product integration. Consequently we define - via the lifting property for $\mathfrak{g}/\mathfrak{h}$ the evolution map $\widetilde{Evol^r} : C^\infty(\mathbb{R}, \mathfrak{g}/\mathfrak{h}) \to C^\infty(\mathbb{R}, \{0\}; G/H, \{H\})$, which is a well defined smooth diffeomorphism.

Next we define the right invariant Lipschitz metrics on $G/H$ via

$$\widetilde{d_\alpha}(gN, hN) := \inf_{n \in N} d_\alpha(gn, h)$$

which provides well-defined right invariant half metrics on $G/H$. They fulfill the Lipschitz-assumptions by definition and the topology is generated by them in the correct sense.

Given finally a $c^\infty$-closed subalgebra $\mathfrak{h}$ of $\mathfrak{g}$, then the set

$$Evol^r(C^\infty(\mathbb{R}, \mathfrak{h}))$$

is a closed subset of $C^\infty(\mathbb{R}, \{0\}; G, \{e\})$, the point evaluations define a subset of $G$, which is a subgroup $H$ by the formula

$$Evol^r(X) \cdot Evol^r(Y)$$
$$= Evol^r(t \mapsto X(t) + Ad(Evol^r(X)(t))Y(t)),$$

which is valid on Frölicher-Lie groups. A curve $c : \mathbb{R} \to H$ is smooth if it is smooth in $G$, so we get the diffeomorphism property. $\qquad\square$

**Remark 7.** *The advantage of this setting is two-fold. Charts are obsolete in this setting and analysis is possible due to Lipschitz-metrics and the Frölicher space structure.*

## References

1. Andreas Kriegl and Peter W. Michor, *The convenient setting of Global Analysis*, Mathematical Surveys and Monographs 53, American Mathematical Society, 1997.
2. Hideki Omori, *Infinite-dimensional Lie groups*, Translation of Mathematical Monographs 158, American Mathematical Society, Providence, Rhode Island, 1997.
3. Josef Teichmann, *Convenient Hille-Yosida Theory*, Revista mathematica complutense, to appear (2002).
4. Josef Teichmann, *A convenient approach to Trotter's formula*, Journal of Lie Theory 11, 400 – 427 (2001).
5. Josef Teichmann, *Infinite dimensional Lie theory from the point of view of functional analysis*, Ph.D. thesis, University of Vienna, 1999, directed by Peter Michor.
6. Josef Teichmann, *Regularity of infinite-dimensional Lie groups by metric space methods*, Tokyo Journal of Mathematics 24, no. 1, 39 – 58 (2001).

# AROUND THE EXPONENTIAL MAPPING

THIERRY ROBART

*Mathematics Department,*
*Howard University, Washington, DC,*
*20059, USA*
*E-mail: trobart@fac.howard.edu*

A major obstacle towards the integrability of infinite dimensional Lie algebras consists in the well-known local non-invertibility of the exponential mapping. Most advances in that direction make use of extrinsic tools reducing the investigation to a laborious case by case analysis. The treatment of the group of smooth diffeomorphisms of a compact manifold $M$ involves for instance a Riemannian connection on $M$. This paper focuses exclusively on intrinsic methods capable of handling the general integration problem. It describes the many ways we can use the exponential mapping itself in order to circumvent its own limitations. The potential of such a canonical approach is illustrated with the class of isotropy Lie algebras of local analytic transformations of $\mathbb{R}^n$, a result recently obtained in collaboration with Niky Kamran.

## 1 Fundamentals of "finite type" Lie groups

*Lie's fundamental discovery was that the complicated nonlinear conditions of invariance of the system under the group transformations could, in the case of a continuous group, be replaced by equivalent, but far simpler, linear conditions reflecting a form of "infinitesimal" invariance of the system under the generators of the group* (Peter Olver [20] )

### 1.1 Lie's discovery

Recall that a geometric structure $\Sigma$ is any subspace of a fiber bundle $B \xrightarrow{\pi} M$ over a finite dimensional manifold $M$ that admits a lifting of the local transformations of $M$. Examples: a Riemannian, Poisson or locally conformal symplectic structure, a system of PDE. Sophus Lie was interested in the local symmetry group $\Gamma$ of such a structure. He discovered, by introducing the associated local Lie algebra $\mathcal{L}$, a linearization method valid for $\Gamma_e$ the connected component of $\Gamma$.

The reconstruction process, passage from the associated linear problem to the global and nonlinear one, is well known in finite dimension. It consists in the exponential mapping followed by translations, and allows one to parametrize $\Gamma_e$ into an analytic manifold compatible with the group operations. We will represent this reconstruction process symbolically by

$$\mathcal{L} \xrightarrow{\text{Exp}} \Gamma_e.$$

The reconstruction process in the case of Lie pseudogroups of infinite type stayed essentially obscure until very recently.

### 1.2 Fundamental theorems

The success of the classical Lie group theory rests essentially on two fundamental theorems respectively called the second (Lie II) and the third (Lie III) fundamental theorems of Lie.

**Theorem 1 (Lie II).** *Let $\mathcal{L}$ be the Lie algebra of a finite dimensional Lie group $G$. Then any Lie subalgebra $\mathcal{H}$ of $\mathcal{L}$ coincides with a unique connected Lie group $H$ imbedded in $G$.*

**Theorem 2 (Lie III).** *Any finite dimensional Lie algebra $\mathcal{L}$ can be integrated into a unique (up to isomorphism) connected and simply connected Lie group $G$.*

## 2 Formal Transformation groups of "infinite type"

It is well known since the sixties (and before)[18,31,7] that the exponential mapping fails outside the Banach setting to provide a local chart and, consequently, a controlled smooth reconstruction process (requested by Lie II or Lie III). This simple fact, sometimes referred to as the "exponential catastrophe", represents the main obstacle towards an infinite dimensional analogue of the theory.

The above mentioned works were all based on formal considerations. The goal of this section is to review the corresponding obstruction[11].

### 2.1 Notations

**Infinitesimal transformations:** Let $\chi(n) = \chi_{-1}(n)$ be the Lie algebra of formal vector fields based at the origin $O$ of $\mathbb{R}^n$. We choose $\{x^1, \ldots, x^n\}$ as coordinates and $\{\partial_1, \ldots, \partial_n\}$ (with $\partial_i = \frac{\partial}{\partial x^i}$) as basis of $\mathbb{R}^n$. For any multi-index $\alpha = (\alpha_1, \ldots, \alpha_n)$ of nonnegative integers we will denote by $|\alpha|$ its degree $\alpha_1 + \cdots + \alpha_n$.

For every non-negative integer $q$, let $\chi_q(n)$ be the Lie subalgebra of formal vector fields tangent to order $q$ to the null vector field. In coordinate form, any formal vector field $V$ of $\chi_q(n)$ decomposes as $V = \sum_{i=1}^{n} V^i \partial_i$ with $V^i = \sum_{|\alpha|>q} v_\alpha^i z^\alpha$. We endow $\chi(n)$ with the natural Tychonov topology.

**Formal transformations:** A formal transformations defined at the origin of $\mathbb{R}^n$ is a formal series

$$\Phi = \sum_{i=1}^{n} (\sum_{|\alpha| \geq 0} \phi_\alpha^i z^\alpha) \partial_i,$$

with an invertible Jacobian $(\phi_j^i)$. We denote by $G(n) = G_{-1}(n)$ the corresponding set of formal transformations in $n$ variables. For any given positive integer $q$, we let $G_q(n)$ be the subset of transformations tangent at order $q$ to the identity transformation $I$.

Thus $G_0(n)$ is the set of formal series

$$\Phi = \sum_{i=1}^{n} (\sum_{|\alpha| \geq 1} \phi_\alpha^i z^\alpha) \partial_i$$

with $\det(\phi_j^i) \neq 0$ $(i,j = 1, \ldots, n)$. The set $G_1(n)$ is the subset of $G_0(n)$ defined by $(\phi_j^i) = I_n$ where $I_n$ is the $n$-by-$n$ identity matrix. Finally for any integer $q \geq 2$, $G_q(n)$ is the set of formal transformations given by

$$\Phi = \sum_{i=1}^{n} (z^i + \sum_{|\alpha| \geq q+1} \phi_\alpha^i z^\alpha) \partial_i.$$

We will write these for short as

$$\Phi = I + \sum_{i=1}^{n} (\sum_{|\alpha| \geq q+1} \phi_\alpha^i z^\alpha) \partial_i.$$

## 2.2   The "exponential catastrophe"

We now analyze the integration problem of the differential equation defining a one-parameter group.

Let $g : (x^i)_{i \in \{1,\ldots,n\}} \rightarrow (X^j)_{j \in \{1,\ldots,n\}}$ be a formal transformation of $G_O(n)$ defined by

$$X^j = \sum_{d=1}^{\infty} \sum_{\substack{i_l \in \{1,\ldots,n\} \\ l \in \{1,\ldots,d\} \\ i_1 \leq \cdots \leq i_d}} a_{i_1 \cdots i_d}^j x^{i_1} \cdots x^{i_d}.$$

Suppose that $g$ preserves the orientation, i.e. that its Jacobian $\det(a_i^j)$ is strictly positive. We would like to know under which condition there exist a unique formal vector field $W = \sum_{k=1}^{n} W^k \partial_k$ with

$$W^k = \sum_{d=1}^{\infty} \sum_{\substack{i_l \in \{1,\ldots,n\} \\ l \in \{1,\ldots,d\} \\ i_1 \leq \cdots \leq i_d}} b_{i_1 \cdots i_d}^k x^{i_1} \cdots x^{i_d},$$

together with an analytic arc

$$\gamma : t \in [0,1] \mapsto \gamma^t$$

of formal transformations satisfying to the differential equation

$$\partial_t \gamma(t, x^1, \ldots, x^n) = W \circ \gamma(t, x^1, \ldots, x^n) \tag{1}$$

with boundary conditions $\gamma^0 = Id$ and $\gamma^1 = g$.

If such a solution exists, the arc $\gamma^t : (x^i)_{i \in \{1,\ldots,n\}} \to (X^j)_{j \in \{1,\ldots,n\}}$ takes the form

$$X^j = \sum_{d=1}^{\infty} \sum_{\substack{i_l \in \{1,\ldots,n\} \\ l \in \{1,\ldots,d\} \\ i_1 \leq \cdots \leq i_d}} \alpha_{i_1 \cdots i_d}^j(t) x^{i_1} \cdots x^{i_d},$$

the coefficients $\alpha_{i_1 \cdots i_d}^j(t)$ being analytic functions of the variable $t$.

For all jet analogue to $\alpha$ we will use the following notation: for all integer $d$ we will denote by $[\alpha]^d$ the matrix

$$(\alpha_{i_1 \cdots i_d}^j)_{\substack{j \in \{1,\ldots,n\} \\ i_l \in \{1,\ldots,n\} \\ l \in \{1,\ldots,d\} \\ i_1 \leq \cdots \leq i_d}}$$

with $n$ rows and $C_{n+d-1}^d$ columns.

By comparing the first order coefficients in equation (1) we get

$$\frac{d}{dt}[\alpha]^1 = [b]^1 \cdot [\alpha]^1, \tag{2}$$

the dot "." denoting the matrix multiplication. Denote by $\mathcal{V}$ the open set in $\text{End}(n, \mathbb{R})$ of matrices with eigenvalues $\lambda_i$ satisfying to $| \Im(\lambda_i) | < \pi$ ($\Im$ stands for the imaginary part). It is a classical result[36] that the exponential mapping of the group $GL(n, R)$ restricted to $\mathcal{V}$ is a local isomorphism. Therefore if $(a_i^j) \in \text{Exp}\mathcal{V}$ the equation (2) admits a unique analytic solution satisfying to the boundary conditions. Let

$$[\alpha]^1(t) = \text{Exp}(t[b]^1)$$

be that solution.

Comparing now the $m$-th order coefficients in (1) we get

$$\frac{d}{dt}[\alpha]^m = [b]^1 \cdot [\alpha]^m + [b]^m \cdot [A^m] + [c]^m \tag{3}$$

where $[A^m]$ is a square matrix of format $C_{n+d-1}^d$ whose coefficients are homogeneous polynomials of degree $m$ depending uniquely on the coefficients of $[\alpha]^1$, and $[c]^m$ is a matrix whose coefficients are polynomials of the coefficients of $[\alpha]^k$ for $1 \le k \le m-1$ and $[b]^k$ for $2 \le k \le m-1$.

Assume now that $[\alpha]^d$ and $[b]^d$ have been obtained by induction for all integer $d \le m-1$. The equation (3) integrates according to

$$[a]^m = [k]^m + e^{[b]^1} \int_0^1 e^{-t[b]^1}[b]^m \cdot [A^m](t)dt$$

where $[k]^m$ is a constant matrix. We observe that, in order for (1) to admit a formal solution with the unique restriction that $(a_i^j) \in \mathrm{Exp}\mathcal{V}$, it is necessary and sufficient that the linear operators

$$\Lambda^m : [b]^m \mapsto \int_0^1 e^{-t[b]^1}[b]^m \cdot [A^m](t)dt$$

be invertible for all integer $m$.

**Theorem 3.** *The formal transformation $g$ admits at least a logarithm if $[b]^1$ has a real spectrum. For all integer $n \ge 2$ the isotropic formal group $G_O(n)$ contains elements as close as desired to the identity transformation that are not in the flow of any vector field.*

*Proof.* We check easily that $\Lambda_{\mathbb{C}}^m$, the complexified of $\Lambda^m$, can be triangularized simultaneously with $[b]^1$; it has as eigenvalues

$$\{\frac{e^{\lambda_{i_1}+\cdots+\lambda_{i_m}-\lambda_k}-1}{\lambda_{i_1}+\cdots+\lambda_{i_m}-\lambda_k}\}_{\substack{1\le i_1\le\cdots\le i_m\le n \\ k\in\{1,\ldots,n\}}}$$

where $\{\lambda_k\}_{1\le k\le n}$ is the spectrum of $[b]^1$. As a result the linear operators $\Lambda^m$ are invertible for all integer $m$. This proves the first part of the theorem. For the seconde part it is enough to consider the bidimensional case . Observe then that the matrix $R_q = \begin{pmatrix} 0 & -\frac{2\pi}{q} \\ \frac{2\pi}{q} & 0 \end{pmatrix}$ admits $\{-\frac{2i\pi}{q}, \frac{2i\pi}{q}\}$ as spectrum. Therefore there exists in the linear group matrices as close as wanted to the identity element for which $\Lambda^m$ is not invertible for all integer $m$. That is in particular the case of $\mathrm{Exp}R_q$ for $q$ large enough integer. This allows to exhibit formal transformations as close as desired to the identity transformation that are not in the image of the exponential map of the isotropy Lie algebra. $\square$

*Remark* 1. The operators $\Lambda^m$ are invertible if and only if the differential $d\text{Exp}_b$ of the exponential mapping at $b$, where $b$ denotes the formal linear vector field of component $[b]^1$, is formally invertible. One can show[9] that

$$\text{Exp}(-b)_\star d\text{Exp}_b = \int_0^1 Ad(\text{Exp}(-tb))dt.$$

This identity is in fact valid for the large class of regular Lie groups (see section 3.1).

*2.3 Illustration*

If $g = \sum_{i=1}^\infty a_i x^i$ is a formal diffeomorphism of $\mathbb{R}$ that fixes $O$ and preserves the orientation (i.e. $a_1 > 0$), its unique formal logarithm $X = \sum_{i=1}^\infty b_i x^i \partial_x$ admits as first terms

$$b_1 = \log(a_1)$$

$$b_2 = -\frac{a_2 \log(a_1)}{a_1 - a_1{}^2}$$

$$b_3 = \frac{2\,a_2\,(a_2 - a_1\,a_2)\,\log(a_1)}{(a_1 - a_1{}^2)\,(a_1 - a_1{}^3)} - \frac{2\,a_3\,\log(a_1)}{a_1 - a_1{}^3}$$

$$b_4 = \frac{a_2\,(-a_2{}^2 + 3\,a_3 - 2\,a_1\,a_3)\,\log(a_1)}{(a_1 - a_1{}^2)\,(a_1 - a_1{}^4)} - \frac{3\,a_4\,\log(a_1)}{a_1 - a_1{}^4}$$

$$+ \frac{\left(2\,a_2 - 3\,a_1{}^2\,a_2\right)\left(\frac{-2\,a_2\,(a_2-a_1\,a_2)\,\log(a_1)}{(a_1-a_1{}^2)(a_1-a_1{}^3)} + \frac{2\,a_3\,\log(a_1)}{a_1-a_1{}^3}\right)}{a_1 - a_1{}^4}$$

Observe that the $b_n$'s are analytic expressions of coefficients $a_1, \ldots, a_n$. The following corollary is immediate.

**Corollary 1.** *Let $g$ be a formal diffeomorphism of the real line. Suppose in addition that $g$ fixes the origin and preserves the orientation; then for all $n$ it admits a unique $n$-th root.*

Let $g^{\frac{1}{2}} = \sum_{i=1}^\infty c_i x^i$ denote the square root of $g = \sum_{i=1}^\infty a_i x^i$; we get as first terms:

$$c_1 = \sqrt{a_1}$$

$$c_2 = \frac{\left(-\sqrt{a_1} + a_1\right) a_2}{(-1 + a_1)\,a_1}$$

$$c_3 = \frac{\frac{-2a_2^2}{(1-a_1)^2} + \frac{4\sqrt{a_1}\,a_2^2}{(1-a_1)^2} - \frac{2a_1a_2^2}{(1-a_1)^2} + \sqrt{a_1}\,a_3}{a_1\,(1+a_1)}$$

$$c_4 = \frac{a_2^3}{(1-a_1)^2} + \frac{a_2^3}{1-a_1} - \frac{5a_2^3}{(1-a_1)^2\,a_1} + \frac{10\,a_2^3}{(1-a_1)^2\,\sqrt{a_1}}$$

$$-\frac{a_2^3}{(1-a_1)\,\sqrt{a_1}} - \frac{12\sqrt{a_1}\,a_2^3}{(1-a_1)^2} + \frac{6\,a_1\,a_2^3}{(1-a_1)^2} + \frac{14\,a_2^3}{(1-a_1)^2\,(1+a_1)}$$

$$+\frac{6\,a_2^3}{(1-a_1)^2\,a_1\,(1+a_1)} - \frac{16\,a_2^3}{(1-a_1)^2\,\sqrt{a_1}\,(1+a_1)} - \frac{4\sqrt{a_1}\,a_2^3}{(1-a_1)^2\,(1+a_1)}$$

$$+3\,a_2\,a_3 - \frac{3\,a_2\,a_3}{1-a_1} + \frac{3\,a_2\,a_3}{(1-a_1)\,\sqrt{a_1}} - 3\sqrt{a_1}\,a_2\,a_3 - \frac{2\sqrt{a_1}\,a_2\,a_3}{1-a_1}$$

$$+\frac{\frac{2a_1a_2a_3}{1-a_1} + \frac{2a_2a_3}{1+a_1} - \frac{3a_2a_3}{\sqrt{a_1}\,(1+a_1)} - \sqrt{a_1}\,a_4 + a_1^2\,a_4}{a_1\,(-1+a_1^3)}$$

## 3   Regularity & pathologies

**Definition 1 (Lie group).** A smooth (respectively analytic) Lie group is a group endowed with a smooth[a] (resp. analytic) manifold structure modeled on a locally convex Hausdorff and complete topological vector space, compatible in the usual sense[b] with the group operations.

Unfortunately such a definition, very fruitful as we know in the finite dimensional context, turns out to be rather powerless in the infinite dimensional one. First mention the lack, outside the Banach setting, of fundamental theorems of differential calculus (inverse function theorem, implicit function theorem and existence and uniqueness theorems for the solution of a Cauchy problem). Secondly, as we will recall later, even in the Banach setting this definition must be refined in order to preserve a certain one-to-one correspondence between Lie algebras and simply connected Lie groups (Lie III).

---

[a] say for the Gâteaux differential calculus[19]
[b] the product and the inversion are smooth (resp. analytic)

### 3.1 A canonical concept

The more promising starting point suggested by Milnor around 1983[19] allows to circumvent to a certain extent the lack of fundamental theorems. It is based on the canonical concept of regularity that we know recall.

Endow the space $C^\infty([0,1],\mathcal{L})$ of smooth paths of the Lie algebra $\mathcal{L}$ of a Lie group $G$ with the $C^\infty$ uniform convergence topology.

**Definition 2 (regular Lie group).** A Lie group is regular whenever the ordinary differential equation[c] $g^{-1}\dot{g}(t) = v(t)$, with $g(0) = e$ as initial condition, admits a smooth solution $\gamma_v$ for all smooth paths $v$ in $\mathcal{L}$ and if the correspondence $v \mapsto \gamma_v(1)$, from $C^\infty([0,1],\mathcal{L})$ with values in $G$, is smooth.

*Remark 2.* The mapping $\mathrm{Exp} : \mathcal{L} \to G$ that associates to the constant path $v$ the group element $\gamma_v(1)$ is nothing but the *exponential map* of the group.

Prior to Milnor's formalization, Omori et al. [22] proposed a strengthened form of regularity. Their viewpoint initially formalized in the Fréchet setting admits a natural extension; we ask in addition for the possibility to construct the solution $\gamma_v$ by "multiplicative integration". This latter is a noncommutative version of Riemann integration. Let us now precise its algorithm.

The multiplicative integral must be conceived as an infinite product of group elements, all infinitesimally close to the identity $e \in G$. Choose a subdivision $0 = t_0 < t_1 < \cdots < t_n = 1$ of the interval $[0,1]$, and, for each segment $[t_{i-1}, t_i]$, a representative $\hat{t}_i \in [t_{i-1}, t_i]$. We have clearly

$$\gamma_v(1) = \prod_{1 \le i \le n} p(t_{i-1})^{-1} \cdot p(t_i).$$

When the subdivision is fine enough, one can approximate $v$ in each segment $[t_{i-1}, t_i]$ by $v(\hat{t}_i)$. This way one expects

$$\prod_{1 \le i \le n} \mathrm{Exp}((t_i - t_{i-1})v(\hat{t}_i))$$

to be close to $\gamma_v(1)$.

Omori et al. request precisely the convergence of this expression when the subdivision mesh tends to zero. The multiplicative integral of $v$ between 0 and 1 is symbolically denoted by

$$\gamma_v(1) = \prod_{0 \le t \le 1} \mathrm{Exp}(v(t)dt).$$

---

[c]the dot denotes the derivative with respect to $t$

**Definition 3 ($\mu$-regular Lie group).** A regular Lie group will be said to be multiplicatively regular (or $\mu$-regular for short) whenever the solution $\gamma_v$ can always be reconstructed by multiplicative integral.

*Remark 3.* All known Lie groups are multiplicatively regular Lie groups! In other words, it is still not known whether or not there exists non-regular Lie groups and regular Lie groups that are not multiplicatively regular.

**Example 1.** The group of smooth contact transformations of a compact manifold $M$ is a nontrivial example[21].

### 3.2 Apparent pathologies

There exists about four types of pathologies that occurred as serious obstructions towards an infinite dimensional analogue of the classical Lie group theory. They are:

(P1) Existence[34] of non-integrable Banach Lie algebras[d];

(P2) Failure[18,4] for the exponential mapping to provide a local chart.

(P3) Existence of topologically complete Lie algebras non-integrable into a group (example: the Lie algebra of local analytic vector fields on the real line);

(P4) Failure for the complexified Lie algebra of a Lie group to be always integrable[16];

This paper focuses only on the two first pathologies.

### 3.3 Concept of S-Lie group

Pathology (P1) has been studied in depth by van Est, Korthagen and Pradines. It is associated to a well-understood cohomological obstruction that we now describe.

cohomological obstruction: Let $\mathcal{L}$ be a Banach Lie algebra with center $\mathcal{Z}$. The Lie algebra $\mathcal{L}$ can be regarded as a central extension of $\mathcal{K}$ by the Lie algebra $\mathcal{Z}$, which is entirely determined by a class $\zeta$ of the second cohomology group (in the sense of Chevalley-Eilenberg[2]) of $\mathcal{K}$ with coefficients in $\mathcal{Z}$.

It is known that $\mathcal{K}$ is integrable into a unique connected and simply connected Banach Lie group $K$. Denote by $H^2(K, \mathcal{Z})$ the second cohomology group (in Eckmann-Eilenberg-MacLane sense[5,6]) of $K$ with coefficients in $\mathcal{Z}$, $K$ acting trivially on $\mathcal{Z}$.

---

[d]see section 3.3

Regard now

$$0 \to \mathcal{Z} \to \mathcal{L} \to \mathcal{K} \to 0 \ (\mathcal{E})$$

not as a Lie algebra extension but rather as an extension of local groups. $(\mathcal{E})$ can be characterized by a natural "germ" of 2-cocycle of a local group $\mathcal{K}_0 \subseteq \mathcal{K}$ with values in $\mathcal{Z}$. This germ defines completely[33] a second class

$$\gamma_{eq}(\mathcal{E}) \in H^2_{eq}(\Gamma_{\mathcal{B}}, \mathcal{Z})$$

of an equivariant cohomology of Vietoris type associated to the local extension $(\mathcal{E})$. Since the equivariant cochain complex is contained in the ordinary one, we have a homomorphism

$$\eta : H^2_{eq}(\Gamma_{\mathcal{B}}, \mathcal{Z}) \to H^2(\Gamma_{\mathcal{B}}, \mathcal{Z}).$$

Put

$$\gamma(\mathcal{E}) = \eta \gamma_{eq}(\mathcal{E}) \in H^2(\Gamma_{\mathcal{B}}, \mathcal{Z}).$$

From the natural identification of $H^2(\Gamma_{\mathcal{B}}, \mathcal{Z})$ with $H^2(K, \mathcal{Z})$, $\gamma(\mathcal{E})$ induces a homomorphism of the second singular homology group (with integer coefficients) of $K$ with values in $\mathcal{Z}$, $\gamma(\mathcal{E}) : H_2(K) \to \mathcal{Z}$. Its image is an abelian subgroup of $\mathcal{Z}$. It is called the *period group* of the local extension $(\mathcal{E})$, and is denoted by $Per(\mathcal{L})$.

**Theorem 4.** *[van Est-Korthagen[34]] A Banach Lie algebra $\mathcal{L}$ is integrable into a Banach Lie group if and only if $Per(\mathcal{L})$ is a discrete sub-group of the centre $\mathcal{Z}$ of $\mathcal{L}$.*

Recovered Lie III: And what if $Per(\mathcal{L})$ is not discrete? Then $\mathcal{L}$ is still integrable into a connected and simply connected topological group that can be interpreted as a quotient of Banach Lie groups. Such a group has a more subtle differential structure comparable to that of a homogeneous leaf space (think, for example, of an irrational winding on the torus - see for instance Douady & Lazard[3] or Serre[28]). This latter can be described by a S-atlas[35].

**Definition 4 (S-Lie group).** A group will be said to be a scheme of Lie group (for short S-Lie group) if it can be interpreted as the quotient of two Lie groups.

**Theorem 5 (Lie III).** *Any Banach Lie algebra $\mathcal{L}$ is integrable into a unique (up to isomorphism) connected and simply connected scheme of Lie group $\Gamma$.*

*Proof.* Denote by $\mathcal{L}_\Pi$ the Banach Lie algebra $C^0([0,1], \mathcal{L})$ of continuous paths of $\mathcal{L}$ with Lie bracket

$$[u, v](t) = [\int_0^t u(\tau)d\tau, v(t)] + [u(t), \int_0^t v(\tau)d\tau],$$

and by $\mathcal{L}_\Lambda$ its ideal that consists in all paths $u : [0,1] \to \mathcal{L}$ satisfying

$$\int_0^1 u(\tau)d\tau = 0.$$

Observe now that $\mathcal{L}_\Pi$ is integrable into a connected and simply connected Banach Lie group $\Pi$ (that turns out to be completely defined on $\mathcal{L}_\Pi$). Let $\Lambda$ be its connected Lie subgroup corresponding to $\mathcal{L}_\Lambda$ (Lie II). Then $\mathcal{L}$ is integrable into the quotient $\Gamma = \Pi/\Lambda$. $\qquad\square$

*Remark* 4. The group $\Pi$ identifies, via the (left or right) logarithmic derivative, with the group of $C^1$-paths in $\Gamma$ that are pointed at the identity while $\Lambda$ is its subgroup of $C^1$-loops. For more about this construction see for instance Swierczkowski[32], Leslie[17] or Robart[27].

*Remark* 5. The period group $Per(\mathcal{L})$ corresponds to the kernel of the exponential map $\mathrm{Exp}_{|\mathcal{Z}} : \mathcal{Z} \subset \mathcal{L} \mapsto \Gamma$.

## 4    Around the exponential rigidity

Despite the "exponential failure" observed in section 2.2 the integration problem of the formal isotropy subgroups of $G_0(n)$ admits an easy and elegant solution. It is based on the classical notion of canonical chart of the second kind[24]. Surprisingly it allows also to recover the "lost" analyticity of a group (see theorem 7 and its following remark). As we will observe in section 5, this concept plays a fundamental role for analytic transformation groups of "infinite type".

### 4.1    Some order

Let $G$ be a Lie group. If $G$ admits the exponential mapping as a manifold chart near the identity we say that $G$ is a Lie group of the *first kind*. The class of Lie groups of first kind is denoted by $\mathcal{EXP}$. If in addition $G$ is analytic, then it is said to be of Campbell-Baker-Hausdorff (CBH for short) type. The class of CBH Lie groups is denoted by $\mathcal{CBH}$. (See Robart[25,26,27] for a complete treatment of Lie II and Lie III in that class.)

We will say that $G$ is a Lie group of second kind if its Lie algebra $\mathcal{L}(G)$ decomposes into a direct sum $\mathcal{L}(G) = \bigoplus_{i=1}^m \mathcal{G}_i$ of vector subspaces $\mathcal{G}_i$ such that the mapping $\prod_{i=1}^m \mathrm{Exp}_i$, that associates to $(x_1, \ldots, x_m) \in \mathcal{G}_1 \times \cdots \times \mathcal{G}_m$ the product $\mathrm{Exp}(x_1) \circ \cdots \circ \mathrm{Exp}(x_m) \in G$, defines a manifold chart near the identity. The integer $m$ denotes the multiplicity of the decomposition.

In this case we will say that $G$ is of the second kind and of order $m$ and belongs to the class $\mathcal{E}XP^m$. We generalize this class to the class $\mathcal{E}XP^{\aleph_0}$ of Lie groups of the second kind and of infinite (countable) order. Moreover, for all $k \in \{1, 2, \ldots, \aleph_0\}$, we will denote by $CBH^k$ the class of analytic Lie groups that belong to $\mathcal{E}XP^k$, so that $CBH^k \subset \mathcal{E}XP^k$. For $m < m'$ we have $CBH \subset \mathcal{E}XP \subset \mathcal{E}XP^m \subset \mathcal{E}XP^{m'} \subset \mathcal{E}XP^{\aleph_0}$.

## 4.2 Formal structure theorems

We are now in position to describe the structure theorems for formal isotropy groups.

**Theorem 6.** *For all integer $n \geq 2$, the formal isotropy group $G_0(n)$ is a Lie group of the second kind that is not of the first kind. Moreover, endowed with their corresponding differential structure of second kind, the formal isotropy Lie groups $G_0(n)$ are all analytic Lie groups, i.e. belong to $CBH^2$*

*Proof.* The fact that $G_0(n)$ is not of the first kind is a consequence of theorem 3. The Lie algebra $\chi_0(n)$ of $G_0(n)$ decomposes as $\text{End}(n, \mathbb{R}) \bigoplus \chi_1(n)$ (for notation see section 2.1). Use this decomposition to exhibit a canonical chart of the second kind and order 2. The analyticity in that chart stems from the Campbell-Baker-Hausdorff character of $\chi_1(n)$ and the analytic action of $\text{End}(n, \mathbb{R})$ on it. □

*Remark 6.* This shows in particular that the inclusion $\mathcal{E}XP \subset \mathcal{E}XP^2$ is strict. In general an extension of a Lie group of class $\mathcal{E}XP^l$ by a Lie group of class $\mathcal{E}XP^m$ gives rise to a Lie group of class $\mathcal{E}XP^{l+m}$.

Let us say that a Lie subgroup $H \hookrightarrow G_0(n)$ of formal transformations is *scalar* whenever the linear part $[b]^1$ of any vector field $W$ of its Lie algebra $\mathcal{H}$ has a real spectrum. Example: the semi-direct product of $T(n, \mathbb{R})$ with $G_1(n)$ where $T(n, \mathbb{R})$ is the linear group of upper-triangular matrices.

**Theorem 7.** *Any scalar isotropy Lie group of formal transformations is a smooth Lie group of the first kind, i.e. belongs to the class $\mathcal{E}XP$. In general, such a Lie group, endowed with its differential structure of first kind, is not analytic; i.e. does not belong to the class $CBH$ of Campbell-Baker-Hausdorff Lie group. Thence the inclusion $CBH \subset \mathcal{E}XP$ is strict.*

*Remark 7.* These scalar Lie groups are analytic when endowed with their second kind differential structure (theorem 6).

*Proof.* The smooth first kind structure is a direct consequence of theorem 3. For the non-analyticity observe that the local analyticity of the Campbell-

Baker-Hausdorff series induces that of the formal series defined by

$$(x,y) \mapsto c(adX)Y = \sum_{k=0}^{\infty} (adX)^k Y.$$

For consider simply $\frac{d}{dt}_{|t=0} H(x, ty)$ (See Robart[27] for more details). Now consider the formal isotropy group $G_0(1)$ of the real line. It is obviously a scalar Lie group. Take $X = -\lambda x \partial_x$ and $Y = x^{p+1} \partial_x$ we get $c(adX)Y = \sum_{i=0}^{\infty} (\lambda p)^i x^{p+1} \partial_x$. We see clearly that for all $\lambda \neq 0$, as small as desired, there is an integer $p$ for which the series $c(adX)Y$ diverges. This completes the proof. $\qquad\square$

## 5   Analytic isotropy transformation groups of "infinite type"

One could be tempted to impute the "exponential failure" solely to the linear obstruction brought out in the formal analysis of section 2.2. It is after all a very strong constraint[14,23,8]. Unfortunately the situation is more subtle. Indeed consider the following: while the formal groups[e] $G_q(n)$ - for $q$ any strictly positive integer - are all exponential none of their subgroups $G_q^\omega(n)$ of analytic transformations is! This 1974 negative result of Jean Écalle [4] left a priori few room for a *canonical* extension of the classical theory. Here is what to do[13,12].

### 5.1   Topology - Notations

Put $M = \mathbb{R}^n$ and fix a norm on $M$. This defines for the space $M \otimes S^p(M^*)$ of homogeneous vector fields of degree $p$ a norm $\| \cdot \|_p$. Recall that a formal vector field $V = \sum_{k=0}^{\infty} V_k$, where, for all $k$, $V_k \in M \otimes S^k(M^*)$, admits a local analytic extension in a neighborhood of the origin in $M = \mathbb{R}^n$ if and only if its coefficients satisfy $\limsup \| V_n \|^{\frac{1}{n}} < \infty$. The class $C^\omega$ of such analytic vector fields does not depend on the chosen norm. Equivalently, a formal series given in coordinate form[f] by

$$V = \sum_{i=1}^{n} \left( \sum_{|\alpha| \geq 0} v_\alpha^i z^\alpha \right) \partial_i$$

---

[e] for notation see section 2.1
[f] see also section 2.1 for notations

is analytic if and only if there exists a positive constant $\rho$ such that for all $i$ and $\alpha$ we have,

$$| v_\alpha^i | \leq \rho^{|\alpha|}.$$

If $\mathcal{L}(\Gamma)$ is any Lie subalgebra of $\chi(n) = \chi_{-1}(n)$ (see section 2.1 for notation), the subspace $\mathcal{L}^\omega(\Gamma) \subset \mathcal{L}(\Gamma)$ of analytic vector fields is an imbedded locally convex and complete topological Lie subalgebra. Let $\rho$ be a positive real number and let $\mathcal{L}_\rho^\omega(\Gamma)$ denote the subspace of $\mathcal{L}^\omega(\Gamma)$ of $V$'s such that $\limsup \| V_n \|_n / \rho^n < +\infty$. We have

$$\mathcal{L}^\omega(\Gamma) = \bigcup_{\rho > 0} \mathcal{L}_\rho^\omega(\Gamma).$$

Each $\mathcal{L}_\rho^\omega(\Gamma)$ is naturally endowed with a Banach space structure of norm

$$\| V \|_\rho = \sup_n \frac{\| V_n \|_n}{\rho^n}.$$

For $\rho < \rho'$ the injection $\mathcal{L}_\rho^\omega(\Gamma) \hookrightarrow \mathcal{L}_{\rho'}^\omega(\Gamma)$ is continuous and compact. This makes $\mathcal{L}^\omega(\Gamma)$ into a complete Hausdorff locally convex topological vector space. Its associated topology is the locally convex strict inductive limit topology

$$\mathcal{L}^\omega(\Gamma) = \varinjlim_{n \in \mathbb{N}} \mathcal{L}_n^\omega(\Gamma).$$

## 5.2  Integrability

Let $\mathcal{L}^\omega(\Gamma) \subset \chi(n)$ be a Lie algebra of local analytic vector fields. For any integer $q$, denote by $\mathcal{L}_q^\omega(\Gamma)$ its subspace of vector fields contained in $\chi_q(n)$. Whenever $q' > q \geq 0$, then $\mathcal{L}_{q'}^\omega(\Gamma)$ is an ideal in $\mathcal{L}_q^\omega(\Gamma)$. Fix now for any integer $q \geq 1$ a linear section $E_q$ of the natural projection

$$\mathcal{L}_{q-1}^\omega(\Gamma) \to \mathcal{L}_{q-1}^\omega(\Gamma)/\mathcal{L}_q^\omega(\Gamma).$$

Clearly the isotropy Lie algebra $\mathcal{L}_0^\omega(\Gamma)$ decomposes as a direct sum of finite dimensional summands

$$\mathcal{L}_0^\omega(\Gamma) = \bigoplus_{q=1}^{+\infty} E_q.$$

We have

**Theorem 8 (Lie III).** *Suppose that we can select the linear sections $E_q$ in the same Banach subspace $\mathcal{L}_\rho^\omega(\Gamma)$ of $\mathcal{L}^\omega(\Gamma)$ (see section 5.1 for notation). Then $\mathcal{L}_0^\omega(\Gamma)$ is integrable into a unique connected and simply connected analytic Lie group $\Gamma_0$ of the second kind and countable order. In other terms $\mathcal{L}_0^\omega(\Gamma)$ is of class $CBH^{\aleph_0}$.*

*Proof.* Let $\mathrm{Exp}_q$ denote the exponential mapping restricted to the vector subspace $E_q$. Simply consider the following left product exponential mapping $\mathrm{PExp}_\lambda$ as candidate for a manifold chart nearby the identity transformation. $\mathrm{PExp}_\lambda$, defined in $\mathcal{L}_0^\omega(\Gamma)$ with values in $\Gamma_0$, associates to the vector field $X = \sum_{k=1}^\infty X_k$ (decomposed with respect to the direct sum $\bigoplus_{q=1}^{+\infty} E_q$) the formal local transformation

$$\mathrm{PExp}_\lambda(X) = \lim_N \mathrm{Exp}_N(X_N) \circ \cdots \circ \mathrm{Exp}_1(X_1).$$

For more details we refer the reader to [13,12] and the references therein. □

**Corollary 2.** *The isotropy group of any Lie pseudogroup $\Gamma$ of analytic transformations (see section 1.1) is a regular analytic Lie group of class $CBH^{\aleph_0}$.*

*Proof.* Its Lie algebra satisfies theorem 8. □

*Remark 8.* This corollary applies in particular to the symmetries of a contact, symplectic or Poisson structure.

We can in fact remove the technical assumption of theorem 8 using an inductive construction. This leads to the following strong extension of the second fundamental theorem of Lie. Recall that a Lie algebra $\mathcal{L}$ is said to be *stable* if it is stable by its adjoint action i.e. $\mathrm{Ad}(\mathrm{Exp}(\mathcal{L}))\mathcal{L} = \mathcal{L}$.

**Theorem 9 (Lie II).** *Any closed and stable Lie subalgebra $\mathcal{H}$ of the Lie algebra $\chi_0^\omega(n)$ of $G_0^\omega(n)$ is integrable into a unique connected subgroup $H$ imbedded in $G_0^\omega(n)$. Such a subgroup $H$ is always a regular analytic Lie group belonging (at least) to the class $CBH^{\aleph_0}$.*

## 5.3 Illustration - Technical remarks

The integration procedure is certainly completely determined as soon as a local chart (or a candidate for a local chart) is fixed. In the sketched proof of theorem 8 that latter depends first on our choice of the direct decomposition $\bigoplus_{q=1}^{+\infty} E_q$ of the Lie algebra. Once this choice is made it becomes also necessary to fix a total order on the set of summands. We are working in a highly non-commutative setting. For instance, for a given direct decomposition we could consider, instead of the left product of exponentials

$$\mathrm{PExp}_\lambda(X) = \lim_N \mathrm{Exp}_N(X_N) \circ \cdots \circ \mathrm{Exp}_2(X_2) \circ \mathrm{Exp}_1(X_1),$$

the right product of exponentials defined by

$$\mathrm{PExp}_\rho(X) = \lim_N \mathrm{Exp}_1(X_1) \circ \mathrm{Exp}_2(X_2) \circ \cdots \circ \mathrm{Exp}_N(X_N).$$

If one chart works so does the other one due to the group stability by the inversion $\iota : g \mapsto g^{-1}$. More generally we expect that any total order will do although it hasn't been proved yet. For the fundamental assumption remains the technical hypothesis explicited in theorem 8.

Illustration: We will now illustrate in the uni-dimensional case the differences between $\mathrm{PExp}_\rho$, $\mathrm{PExp}_\lambda$ that provide local charts and the exponential mapping Exp that fails to be a chart. Surprisingly these three maps look very similar from a formal viewpoint.

Consider $\chi_1^\omega(1)$ the Lie algebra of analytic vector fields on the real line that are tangent to order 1 with the null vector field. It consists in the analytic vector fields of the form $X = \sum_{i=1}^\infty a_i x^{i+1} \partial_x$. If $g = x + \sum_{i=1}^\infty g_i x^{i+1}$ represents its exponential then it admits as first terms

Exponential mapping:

$$g_1 = a_1$$

$$g_2 = a_1^2 + a_2$$

$$g_3 = a_1^3 + \frac{5}{2} a_1 a_2 + a_3$$

$$g_4 = a_1^4 + \frac{13}{3} a_1^2 a_2 + \frac{3}{2} a_2^2 + 3 a_1 a_3 + a_4$$

$$g_5 = a_1^5 + \frac{77}{12} a_1^3 a_2 + \frac{35}{6} a_1 a_2^2 + 6 a_1^2 a_3 + \frac{7}{2} a_2 a_3 + \frac{7}{2} a_1 a_4 + a_5$$

Decompose now $\chi_1^\omega(1)$ using the homogeneous factors as summands. This leads to the following charts for the group $G_1^\omega(1)$ of analytic transformations of the real line tangent to order 1 to the identity.

Right product exponential: It associates to $X = \sum_{i=1}^{\infty} a_i x^{i+1} \partial_x$ the group element

$$\mathrm{PExp}_\rho(X) = \lim_n \mathrm{Exp}(a_1 x^2 \partial_x) \circ \mathrm{Exp}(a_2 x^3 \partial_x) \circ \cdots \circ \mathrm{Exp}(a_n x^{n+1} \partial_x).$$

Represented by $r = x + \sum_{i=1}^{\infty} r_i x^{i+1}$ we get as first terms

$$r_1 = a_1$$

$$r_2 = a_1{}^2 + a_2$$

$$r_3 = a_1{}^3 + 2\, a_1\, a_2 + a_3$$

$$r_4 = a_1{}^4 + 3\, a_1{}^2\, a_2 + \frac{3}{2}\, a_2{}^2 + 2\, a_1\, a_3 + a_4$$

$$r_5 = a_1{}^5 + 4\, a_1{}^3\, a_2 + 4\, a_1\, a_2{}^2 + 3\, a_1{}^2\, a_3 + 3\, a_2\, a_3 + 2\, a_1\, a_4 + a_5$$

Left product exponential: It associates to $X = \sum_{i=1}^{\infty} a_i x^{i+1} \partial_x$ the group element

$$\mathrm{PExp}_\lambda(X) = \lim_n \mathrm{Exp}(a_n x^{n+1} \partial_x) \circ \cdots \circ \mathrm{Exp}(a_2 x^3 \partial_x) \circ \mathrm{Exp}(a_1 x^2 \partial_x).$$

Represented by $l = x + \sum_{i=1}^{\infty} l_i x^{i+1}$ we get as first terms

$$l_1 = a_1$$

$$l_2 = a_1{}^2 + a_2$$

$$l_3 = a_1{}^3 + 3\, a_1\, a_2 + a_3$$

$$l_4 = a_1{}^4 + 6\, a_1{}^2\, a_2 + \frac{3}{2}\, a_2{}^2 + 4\, a_1\, a_3 + a_4$$

$$l_5 = a_1{}^5 + 10\, a_1{}^3\, a_2 + \frac{15}{2}\, a_1\, a_2{}^2 + 10\, a_1{}^2\, a_3 + 4\, a_2\, a_3 + 5\, a_1\, a_4 + a_5$$

It is interesting to compare the three series. Observe first that in all cases the coefficients $c_n$ of $x^{n+1}$ are polynomials in the $a_1, a_2, \ldots$ variables involving only monomials $a_{i_1}^{p_1} a_{i_2}^{p_2} \cdots a_{i_l}^{p_l}$ satisfying $p_1 i_1 + p_2 i_2 + \cdots + p_l i_l = n$. Second, and for obvious reasons, the coefficients of these monomials are always positive rational numbers. Finally and strikingly there exists a "natural order" between these three series. Indeed observe that the corresponding coefficients for the right product exponential are always smaller than those of the exponential mapping that, in turns, are smaller than those of the left product exponential. For example if we focus on the coefficient of $a_1^2 a_3 x^6$ we get 3 for $PExp_\rho$, 6 for Exp and 10 for $PExp_\lambda$. This is no accident; our observation remains valid at any level of the three series! We recast it as follows.

**Definition 5.** A series $\Sigma$ in the variables $x, a_1, a_2, \ldots$ will be said to be positive, and we will write $\Sigma \geq 0$, if each of its monomials is affected by a positive (nonnegative) factor. If $\Sigma_1$ and $\Sigma_2$ are two such series, then we will say that $\Sigma_1$ is smaller than $\Sigma_2$ and write $\Sigma_1 \leq \Sigma_2$ whenever $\Sigma_2 - \Sigma_1 \geq 0$. If in addition $\Sigma_2 - \Sigma_1$ is not identically zero then we write $\Sigma_1 < \Sigma_2$.

**Theorem 10.** *The previous series satisfy*

$$0 < PExp_\rho < Exp < PExp_\lambda.$$

*Proof.* The proof relies on Lie's series for the exponential mapping

$$\mathrm{Exp}(X)(x) = x + \sum_{n=1}^{+\infty} \frac{1}{n!} X^n(x)$$

and the fact that, if $p < q$, the Lie derivative of $x^p$ with respect to $x^q \partial_x$, precisely equal to $px^{p+q-1}$, is always smaller than the Lie derivative of $x^q$ with respect to $x^p \partial_x$. $\qquad\square$

*Remark* 9. Theorem 10 shows, if necessary, to which extent the analysis in this infinite dimensional context is *extremely* delicate. Bearing in mind the non-invertibility of the exponential mapping Exp, this "sandwich" theorem gives us little hope to reach the local invertibility of $PExp_\lambda$ or $PExp_\rho$ using an iteration method of Newton-Kantorovich type.

*Remark* 10. Theorems 8 & 9 suggest strongly the possibility to handle in the infinite dimensional context the reconstruction process (see section 1.1) by using locally and systematically products of exponentials. In particular *any* *basis* of an integrable Lie algebra should provide, in principle, a family of local charts.

**References**

1. E. Cartan, *Les systèmes différentielles extérieurs et leur applications géométriques*, (Hermann, 1945).

2. C. Chevalley; S. Eilenberg: *Cohomologie of Lie groups and Lie algebras*, Trans. A.M.S. **63**, (1948), 85-154.
3. A. Douady and M. Lazard, *Espaces fibrés en algèbres de Lie et en groupes*, Inventiones Mathematicae 1 (1966), 133-151.
4. J. Ecalle, *Théorie itérative: introduction à la théorie des invariants holomorphes*, J. Math. Pures Appl. **54** (1975), 183-258.
5. Eckmann,B. *Der Cohomologie-Ring einer beliebigen Gruppe*, Comm. Math. Helv. 18, (1946), 232-282.
6. Eilenberg,S.; MacLane,S. *Cohomology theory in abstract groups. I*, Ann Math Vol.**48**, No.1, January, (1947), 51-78.
7. C. Freifeld, One-parameter subgroups do not fill a neighborhood of the identity in an infinite-dimensional Lie (pseudo-) group, Battelle Rencontres, 1967 Lectures in mathematics and physics (Benjamin, New-York), 538-543.
8. J. Grabowski, Free subgroups of diffeomorphism groups. Fundamenta Mathematicae , **131**, (1988), 103-121.
9. J. Grabowski, Derivative of the exponential mapping for infinite dimensional Lie groups, Annals Global Anal. Geom. **11** (1993), 213-220.
10. V. Guillemin and S. Sternberg, *An algebraic model of transitive differential geometry*, Bull. of the A.M.S. **70** (1964), no. 1, 16–47.
11. N. Kamran, T. Robart, Sur les pseudogroupes abstraits de type F, Canadian Mathematical Society, Conference Proceedings, Volume **21**, 1997.
12. N. Kamran, T. Robart, On the parametrization problem of Lie pseudogroups of infinite type, **C. R. Acad. Sci. Paris**, Série I, t.331 (2000), p.899-903.
13. N. Kamran, T. Robart, A manifold structure for analytic isotropy flat Lie pseudogroups of infinite type, **Journal of Lie Theory**, vol. **11** (2001), Number 1, p.57-80.
14. N. Kopell, Commuting diffeomorphisms, Proc. Symp. Pure Math. Amer. Math. Soc. **14** (1970), 165-184.
15. M. Kuranishi, *Lectures on involutive systems of partial differential equations*, Publ. Soc. Mat. Säo Paulo, 1967.
16. L. Lempert, *The problem of complexifying a Lie group*, Proceedings of "Multidimensional complex analysis and partial differential equations" (So Carlos, 1995), vol. 205, pp. 169–176, Amer. Math. Soc., Providence, RI, 1997.
17. J. Leslie, *On the integrability of some infinite dimensional Lie Algebras* preprint, Howard University (Washington DC).
18. D. Lewis, *Formal power series transformations*, Duke Math. J. **5** (1939), 794–805.

19. J. Milnor, *Remark on infinite-dimensional Lie groups*, Proceedings of les Houches' Summer School, session XL, pp. 1007–1057, North-Holland, 1984.

20. P. J. Olver, *Applications of Lie groups to differential equations*, Graduate texts in mathematics, vol. 107, Springer-Verlag, New York, 1993.

21. H. Omori, *Infinite dimensional Lie transformations groups*, Lecture Notes in Math., vol. 427, Springer-Verlag, 1974.

22. H. Omori; Y. Maeda; A. Yoshioka; O. Kobayashi: *On regular Fréchet-Lie groups*. Tokyo J.Math. vol. 5, No.2, (1981), 365-397.

23. J. Palis, *Vector fields generate few diffeomorphisms*, Bull. of A.M.S. Vol.80, No.3, (1974), 503-505.

24. L. Pontrjagin, *Topological groups*, Princeton University Press, fifth printing, 1958.

25. T. Robart, *Groupes de Lie de dimension infinie. second et troisième théorèmes de Lie. i - Groupes de première espèce*, C. R. Acad. Sci. Paris, Série I **322** (1996), 1071–1074.

26. T. Robart, *Sur l'intégrabilité des sous-algèbres de Lie en dimension infinie*, Can. J. Math., vol. **49** (4), 1997 pp. 820-839.

27. T. Robart, *On Milnor's regularity and the path functor for the class of infinite dimensional Lie algebras of CBH type*, Algebras, Groups and Geometries **16**, 1998.

28. J.P. Serre, *Lie algebras and Lie groups*, Lecture Notes in Math. **1500**, Springer-Verlag, 1965.

29. I.M. Singer, S. Sternberg, *The infinite groups of Lie and Cartan, Part I (the transitive groups)*, J. d'Anal. Math. 15 (1965) , 1–114.

30. S. Sternberg, *On the structure of a local homeomorphism*, Amer. J. Math. 80 (1958) , 623–631.

31. S. Sternberg, *Infinite Lie groups and the formal aspects of dynamical systems* J. of Math. and Mech., vol. 10, No. 3 (1961), 451-474.

32. S. Świerczkowski, *The path-functor on Banach Lie algebras*, Proc. Konink. Nederl. Akad. Wetensch., Ser A-74, N⁰3 (1971), 235-239.

33. Est, W.T. van. *Local and global groups I*, Nederl Akad Wetensch Proc Ser A-65 Indagationes Math. **24**, (1962), 391-425.

34. W.T. van Est and T.J. Korthagen, *Nonenlargeable lie algebras*, Proc. Konink. Nederl. Akad. Wetensch., Ser A-67=Indag. Math. **26** (1964), 15–31.

35. W.T. van Est, *Rapport sur les S-atlas*, Astérisque **116** (1984), 235–292.

36. V.S. Varadarajan, *Lie groups, Lie algebras, and their representations*; Graduate texts in mathematics, vol. 102, Springer-Verlag, New York, 1984.

# ON A SOLUTION TO A GLOBAL INVERSE PROBLEM WITH RESPECT TO CERTAIN GENERALIZED SYMMETRIZABLE KAC-MOODY ALGEBRAS

JOSHUA A. LESLIE

*Mathematics Department,*
*Howard University, Washington, DC,*
*20059, USA*
*E-mail: jleslie@fac.howard.edu*

Given a generalized symmetrizable Kac-Moody algebra we show how to construct a diffeological Lie group whose Lie algebra is a completed version of the said Kac-Moody algebra.

## Introduction

In the case of infinite dimensional topological Lie algebras, it is well known that there exists Lie algebras which are not realizable as the Lie algebra of right invariant vector fields on a smooth manifold group. Since the advent of Kac-Moody algebras it has been natural to ask if there are groups associated to these Lie algebras in such a way that the representation theory of the Lie algebras are related to the representation theory of the associated groups in analogy with the theory of finite dimensional Lie groups. Goodman and Wallach [6] succeeded in associating Banach-Lie groups to affine Kac-Moody algebras and used them to study the representation theory of affine Kac-Moody algebras.

In these notes we shall assign a Hausdorff locally convex topology to a generalized Kac-Moody algebra, $\mathfrak{G}(A)$, in such a way that the Lie algebra becomes a topological algebra. We then show how to construct a Lie group structure on $C^\infty(I, \bar{\mathfrak{G}}(A))$, where $\bar{\mathfrak{G}}(A)$ is an appropriate completion of a generalized Kac-Moody algebra associated to a symmetrized generalized Cartan matrix A, where $C^\infty(I, \bar{\mathfrak{G}}(A))$ is the space of $C^\infty$ functions from the unit interval I into $\bar{\mathfrak{G}}(A)$.

Further, we exhibit an exact sequence

$$0 \to \Omega \to C^\infty(I, \bar{\mathfrak{G}}(A)) \to \bar{\mathfrak{G}}(A) \to 0$$

of smooth Lie algebra homomorphisms, where

$$\Omega = \{f \in C^\infty(I, \bar{\mathfrak{G}}(A)) : \int f dt = 0\};$$

31

in the case of the above exact sequence, $C^\infty(I, \tilde{\mathfrak{G}}(A))$ also designates by abuse of notation the Lie algebra of the above mentioned Lie group $C^\infty(I, \tilde{\mathfrak{G}}(A))$ (the underlying topological structures are the same (see [13] for details)). There is a normal subgroup, $\omega$, of $C^\infty(I, \tilde{\mathfrak{G}}(A))$ having $\Omega$ as its Lie algebra. We shall show that $C^\infty(I, \tilde{\mathfrak{G}}(A))/\omega = \Lambda(A)$ in the category of diffeological groups has many of the requisite properties for a Lie group corresponding the canonical diffeological Lie algebra structure on $\tilde{\mathfrak{G}}(A)$.

Finally, we lay the groundwork to show that the highest weight representations of $\tilde{\mathfrak{G}}(A)$ are differentials of smooth representations of $\Lambda(A)$ in the category of diffeological modules.

In section 1 we redefine diffeological spaces and reformulate some of the results on them that we subsequently make use of. Most of the verifications follow easily from the work done by Souriau, Dazord, Donato, Isglesias (e.g. see [3], [4], [7],[22]). In those cases where we state formerly unpublished results on diffeological spaces or diffeological groups we give proofs or indications of proofs when it seems appropriate. The diffeological spaces that we deal with are different from the differential spaces considered by the Polish school of Sikorski et al (see e.g. [20]); there, the generalization of manifolds studied is in terms of the commutative ring of smooth functions defined on a manifold. Here, we are looking instead at a generalization of the structure given by the smooth functions defined with domain an open subset of an arbitrary Hausdorff, complete, locally convex topological vector space and with range a fixed manifold.

In section 2 we give details on our construction of a diffeological Lie group $\Lambda(A)$, corresponding to a generalized symmetrized Cartan matrix A with a possibly countably infinite number of rows and columns such that the rows and columns are uniformly $\ell_2$ bounded. Our definition of generalized Kac-Moody Lie algebra is more general than the one discussed by Kac in his most recent edition of Infinite Dimensional Lie Algebras, but less general than Borcherds' [1] in that we suppose that the Cartan subalgebra has a canonical Hilbert space structure.

## 1   Diffeological algebraic structures

Let $\Gamma$ be the Grassmannian ring of super numbers generated by an arbitrary set $\mathcal{X} = \{x_i\}_{i \in I}$ with its topology given by $\mathrm{indlim}_{i \in J}\Gamma_i$, where $J$ is the collection of finite subsets of $I$ ordered by inclusion and $\Gamma_i$ the finite dimensional subspaces of $\Gamma$ generated by $x_{i_1}, ..., x_{i_{n_i}}$. With this topology $\Gamma$ is a complete locally convex topological vector space. $\Gamma_i$ is a $Z_2$ graded commutative (i.e. $ab = (-1)^{|a||b|}ba$ algebra, where $|a|$ designates the parity of a. $\Gamma$ with this topology will be used as our base ring throughout what follows.

Let V and W be a topological graded modules over $\Gamma$, a continuous mapping

$$f : V \times \cdots \times V \mapsto W$$

is said to be an $n - multimorphism$ when f is n-multilinear with respect to the ground field $\overline{K}$ and

$$f(e_1, \ldots, e_i\gamma, e_{i+1}, \ldots, e_n) = f(e_1, \ldots, e_i, \gamma e_{i+1}, \ldots, e_n), \gamma \in \Gamma$$

and

$$f(e_1, \ldots, e_n\gamma) = f(e_1, \ldots, e_n)\gamma, \gamma \in \Gamma.$$

Now suppose $U \subseteq V$ open, and suppose W is a second topological $\Gamma$ module. A function f:U$\mapsto$W will be called super $C^n$ or $G^n$ or simply smooth when there exists continuous maps, which are k-multimorphisms in the k-terminal variables, for x$\in$U fixed, $D^k f(x; \cdots) : U \times V \times \cdots \times V \mapsto W, k \leq n$, such that, for $1 \leq k \leq n$,

$$F_k(h) = f(x + h) - f(x) - 1/1!Df(x, h) - \cdots - 1/k!D^k f(x, h, \cdots, h)$$

satisfies the property that

$$G_k(t, h) = \begin{cases} F_k(th)/t^k, & t \neq 0 \\ 0, & t = 0 \end{cases} \tag{1}$$

In [22] Souriau introduces the notion of a Diffeological Space. In this paper we shall reformulate his definition to include infinite dimensional supermanifolds among the canonical domains in the definition of a diffeological structure.

Let $\mathfrak{M}$ be the category whose objects are super manifolds modeled on the open subsets of graded complete locally convex topological vector spaces (chlctvs) (see [12]); the morphisms of $\mathfrak{M}$ are the $C^\infty$ functions. A diffeological space is a set S together with a contravariant subfunctor $F_S(C) \subseteq Hom_{set}(C, S)$ such that constant maps are in $F(C)$ for each object C of $\mathfrak{M}$ and each $x \in S$ and such that F restricted to the subcategory of open subsets of a fixed super manifold, C, whose morphisms are the canonical injection of open subsets of C into each other satisfies the axioms of a set valued sheaf.

When S is a $C^\infty$ super manifold we shall suppose without explicit mention to the contrary that it has its underlying diffeology given by $F(C)$ being defined to be the set of $C^\infty$ maps with domain C and values in S.

Given a super manifold M and a diffeological space (S, F), a mapping f from a subset $C \subseteq M$ to S is called smooth when there exists an open neighborhood U of C and a smooth extension $\tilde{f}$ of f to U, $\tilde{f} : U \mapsto S$.

Given any collection of diffeological structures on a set S, $F_i$, we have that $\cap F_i$ is a diffeological structure, thus any assignment of functions, $G_S(U) \subset Hom_{set}(U,S), U \subset E$, E a Hausdorff locally convex topological vector space, generates a diffeology; namely, the smallest or finest diffeology containing the $G_S(U)$. For the diffeology so generated we shall call $G_S(U)$ a system of generators.

A useful notion for diffeological structures is that of the pull-back; given a diffeological structure on a set T, (T, G), and a function, $f : S \mapsto T$, define $f^*G(U) = \{g \in Hom_{sets}(U,S) : f \circ g \in G(U)\}$. It is straightforward to verify that $f^*G$ is a diffeological structure on S.

Given a subset $S_1 \subset S_2$, where $(S_2, F_2)$ is a diffeological space, there is a diffeological structure induced on $S_1$ by $F_1(C) = \{f \in F_2(C) : f(C) \subset S_1\}$. Note that $F_1 = i^*(F_2)$, where $i : S_1 \mapsto S_2$ is the canonical inclusion. In the rest of this paper when we consider a subset of a diffeological space as a diffeological space it will be with the above described structure unless there is an explicit mention to the contrary. When C is an open subset of a graded hlctvs we shall call f of $F_S(C)$ a plot of the diffeological structure at a point s = f(x), $x \in C$.

**Definition 1.1.** We shall call a diffeological structure _lattice_ or L type when given two plots $f : M_1 \mapsto S$, $g : M_2 \mapsto S$ at a point f(x) = g(y) there always exists a third plot through which the germs of f at x and g at y factor; that is, there exists a plot $h : N \mapsto S$ such that $f = h \circ \phi, g = h \circ \gamma$, where $\phi : \tilde{M}_1 \mapsto N, \gamma : \tilde{M}_2 \mapsto N$ are smooth functions such that $\phi(x) = \gamma(y)$, where $\tilde{M}_i \subset M_i$ is a neighborhood of x (resp. y).

Given diffeological spaces $(X_1, F_1)$ and $(X_2, F_2)$ a function $f : X_1 \mapsto X_2$ is called smooth or $C^\infty$ for each $g \in F_1$, we have $f \circ g \in F_2$. We shall say that f is locally smooth at $x_0 \in X_1$ when given any smooth map, $g \in F_1$, from a neighborhood, U, of $0 \in E$, where $f(0) = x_0$ there exists a neighborhood, $U_0 \subset U$ of 0 such that $f \circ (g|U_0) \in F_2$.

Given diffeological spaces $(X_1, F_1)$ and $(X_2, F_2)$, and an open subset U of a graded chlctvs E a mapping $f : U \mapsto C^\infty(X_1, X_2)$ is _smooth_ when $F : U \times X_1 \mapsto X_2$, given by F(u, x) = f(u)(x) is smooth. The finest diffeology on $C^\infty(X_1, X_2)$ admitting these functions as generating plots will be called the function space diffeology on $C^\infty(X_1, X_2)$.

¿From the definitions, using the pull-back, we get the following:
Suppose that (S, F) is a diffeological space of L-type; consider the equivalence relation generated between germs of one dimensional plots at $s \in S$ $C_1$ and

$C_2$ are open intervals containing 0 as follows: let f and g be one dimensional plots at 0 with domains open intervals $C_1$ and $C_2$ containing 0, we write $f_0 \equiv g_0$, when f(0) = g(0) = s and there exists a plot k:$U \mapsto S$ through which the germs of f and g factor at 0 and we have that $D_0(h \circ f) = D_0(h \circ g)$, where h is any smooth real valued function defined on the image of k. The equivalence classes will be called tangent vectors at s, we shall designate the set of tangents at s by $T_s S$. It is immediate that a smooth map $f : S \mapsto W$ defines a function $Tf : T_s S \mapsto T_{f(s)} W$. When M is a manifold modeled on a locally convex topological vector space, this definition is equivalent to the classical one. Given a plot $f : U \mapsto S$, we use $Tf : TU \mapsto TS$ to define plots on TS and thus a diffeology on TS, in what follows TS will be considered as a diffeological space with the finest diffeology admitting such Tf's as plots.

Given diffeological spaces $(X_1, F_1)$ and $(X_2, F_2)$ the Cartesian product diffeology is defined by $f \in F_1 \times F_2$ if and only if $pr_1 \circ f \in F_1$ and $pr_2 \circ f \in F_2$. A group G with a diffeology $\mathfrak{F}$ on its underlying set will be called a diffeological group when multiplication and inversion define smooth maps $G \times G \mapsto G$ and $G \mapsto G$; in a similar vein we define the notions of diffeological vector space and diffeological Lie algebra.

**Definition 1.2.** A diffeological vector space E will be called <u>integral</u> when there exists a smooth linear map $\int : C^\infty(I, E) \mapsto E$, such that given any smooth real valued linear function $H : E \mapsto R$, we have $H(\int(f)) = \int H(f(t))dt$ and such that given any $v \in E$ we have $\int(f(t)v) = (\int f(t)dt)v$, for any smooth real valued function on the unit interval.

A subspace, K, of an integral diffeological vector space, E, will be called <u>closed</u> when $\int(C^\infty(I, K)) \subseteq K$.

One readily verifies that each diffeological group is of L-type.

Given a plot at the identity, $e \in G, f : U \mapsto G, x \in U, U \subset E$, E a locally convex topological vector space, U open, define $Df(x; \alpha) = [f(x + t\alpha)]_{t=0}$, for $x \in U$.

We have

**Proposition 1.1.** *If G is a diffeological group, then $T_e G$ is a diffeological vector space.*

and

**Proposition 1.2.** *Let G be a diffeological group, then TG is a diffeological group, where the group operation is given by $TP \circ \sigma^{-1}$, where $\sigma : T(G \times G) \mapsto TG \times TG$ is the canonical map, and $P : G \times G \mapsto G$ is the diffeological group operation.*

The proofs of the above propositions on the diffeological group structure of TG are to be found in [14].

*Remark 1.1.* Let G be a diffeological group and N a normal subgroup of G, then one has that G/N is a diffeological group.

Before proceeding we shall adapt what we need of Iglesias's treatment [6] of diffeological fiber bundles to our context.

Given diffeological spaces (X,G) and (Y, K) and a smooth map $f : X \mapsto Y$ consider the diffeological groupoid of automorphisms of f, $\mathfrak{G}_f$, it has as objects the elements of Y, its morphisms, $Morph_f(y, y')$, are the difffeomorhisms from $f^{-1}(y)$ to $f^{-1}(y')$. Let $s : Morph_f \mapsto Y$ (resp. $t : Morph_f \mapsto Y$ be the source map (resp. target map); that is, given $h \in Morph_f(y, y')$, s(h) = y (resp. t(h) = y'). We now define a diffeology on $\mathfrak{G}_f$ as follows: given an open subset U of a graded hlctps E, we say that $g : U \mapsto Morph_f$ is a plot if and only if we have that

i) $s \times t \circ g$ is smooth,

and

ii) with $X_s :\equiv \{(x, y) : x \in U, y \in y_x\} \subset U \times X$, the mapping $\rho : X_s \mapsto X$, given by $\rho(x, y) = g(x)(y)$, is smooth,

and

iii) setting $X_t :\equiv \{(x, y) : x \in U, y \in y'_x)\} \subset U \times X$, we have that the mapping $\mu : X_t \mapsto X$ given by $\mu(x, y) = g(x)^{-1}(y)$ is smooth.

We take on $\mathfrak{G}_f$ the finest diffeology which induces the given diffeology on Y and admits the above maps to the morphisms of $\mathfrak{G}_f$ as smooth maps. This diffeology on $\mathfrak{G}_f$ will be called the standard diffeology.

It follows in a straightforward manner from the definitions that

**Proposition 1.3.** *The standard diffeology endows $\mathfrak{G}_f$ with the structure of a diffeological groupoid; that is, setting $S_{def} :\equiv \{(x, y) \in Morph \times Morph : t(x) = s(y)\} \subset Morph \times Morph, \sigma(x, y) :\equiv y \circ x$ defines a smooth map $\sigma : S_{def} \mapsto Morph$, the inverse operation in the groupoid defines a smooth map $Morph \mapsto Morph$, and $\iota(m) = i_m : f^{-1}(m) \mapsto f^{-1}(m)$ defines a diffeomorphism from the objects of $\mathfrak{G}$ to the identities of $\mathfrak{G}_f$.*

It is important to note that the following fundamental property of vector bundles on Lie groups is satisfied for diffeological groups (see [14]).

**Theorem 1.1.** *Let G be a diffeological group and let $\Pi : TG \mapsto G$ be the canonical map, then $s \times t : Morph_\Pi \mapsto G \times G$ satisfies the local lifting property; that is, given any open subset of a graded hlctvs $U \subset E, x \in U$ and smooth map $f : U \mapsto G \times G$, there exists an open neighborhood $U_0 \subset U$ of x and smooth map $F : U_0 \mapsto Morph_\Pi$ such that $s \times t \circ F = f$.*

**Definition 1.3.** A diffeological group G will be called a diffeological Lie group when the tangent space at the identity, $T_e G$, is a diffeological vector space which

i) admits for every $\alpha \in T_e G$ a smooth real valued linear map $T : T_e G \mapsto R$ such that $T(\alpha) \neq 0$.

and for which

ii) the linear plots of $T_eG$ are cofinal in the sense that every plot of $T_eG$ factors smoothly through a smooth linear map of a complete Hausdorff locally convex topological vector space into $T_eG$.

**Theorem 1.2.** *Let $G$ be a diffeological Lie group, then $T_eG$ admits the structure of a Lie algebra such that the bracket operation defines a smooth linear map $\nabla_X : T_eG \mapsto T_eG$, where $\nabla_X(Y) = [X, Y]$.*

Let $E$ be an Hausdorff, sequentially complete, locally convex topological vector space, we suppose $E$ furnished with the canonical diffeology, and let $\text{Aut}(E)$ be the group of linear diffeomorphisms of $E$. Given an arbitrary locally convex topological vector space $V$ and an open subset $U \subset V$, a mapping $f : U \mapsto \text{Aut}(E)$ is a plot iff i) $F : U \times E \mapsto E$ defined by $F(x,y) = f(x)(y)$ is smooth and ii) $G : U \times E \mapsto E$ defined by $G(x, y) = f(x)^{-1}(y)$ is smooth.

The finest diffeology on $\text{Aut}(E)$ admitting the above plots of $\text{Aut}(E)$ will be called the normal diffeology on $\text{Aut}(E)$.

**Theorem 1.3.** *$\text{Aut}(E)$ with the normal diffeology is a diffeological Lie group.*

**Proposition 1.4.** *Let $G$ be a regular diffeological Lie group, then there exists a smooth function $\exp: T_eG \mapsto G$ such that $\exp((t + s)) = \exp(t\xi) \times \exp(s\xi)$ and $[\exp(t\xi)]_t = R_{\exp(t\xi)}(\xi)$.*

**Corollary 1.1.** *If $G$ is a regular diffeological Lie group, then $T_eG$ admits the structure of a diffeological Lie algebra.*

**Definition 1.4.** *Let $\mathfrak{L}$ be an integral diffeological Lie algebra, $\mathfrak{L}$, A closed ideal, $\mathfrak{J}$, will be called pre-integrable when for each i) $a(t) \in C^\infty(I,\mathfrak{L})$ the differential equation, $(\star)$ $y' = [a(t), y]$, admits a smooth flow, $\Phi(a(t), s, l)$ such that $D_s\Phi(a(t), s, l) = [a(s), \Phi(a(t), s, l)]$, $\Phi(a(t), 0, l) = l$ and defines a smooth map $\Phi : C^\infty(I,\mathfrak{J}) \times I \times \mathfrak{J} \mapsto \mathfrak{J}$, ii) further, we suppose that $\Phi(a(t), s, \cdot) : \mathfrak{J} \mapsto \mathfrak{J}$ is a diffeomorphism which defines a diffeomorphism $\Phi : C^\infty(I,\mathfrak{J}) \mapsto C^\infty(I,\mathfrak{J})$, where $\Phi(f)(t) = \Phi(a(t), t, f(t))$*

Lie's second fundamental theorem takes the form:

**Theorem 1.4.** *Let $G$ be a regular Lie group and $\mathfrak{H}$ a pre-integrable diffeological Lie subalgebra of $T_eG$ with an ideal, $\mathfrak{K}$, as a diffeological vector space complement; that is, a mapping, $f : U \mapsto T_e\vec{G} = H \times K$ is smooth if and only if $\pi_H \circ f$ and $\pi_K \circ f$ are smooth. Then there exists a regular diffeological Lie subgroup, $H$, of $G$ such that the canonical injection, $i : H \mapsto G$ induces an isomorphism of diffeological Lie algebras, $T(i) : T_e H \mapsto \mathfrak{H} \subset T_eG$.*

**Theorem 1.5.** *Let $G$ be a simply connected, regular, diffeological Lie group with canonical diffeomorphism $\chi : C_0^\infty(I, G) \mapsto C^\infty(I, T_e G)$ and suppose that $H$ is a connected normal subgroup of $G$ such that there exists a pre-integrable Lie ideal, $\mathfrak{H} \subseteq T_eH \subseteq T_eG = \mathfrak{G}$, of the Lie algebra, $\mathfrak{G}$, of $G$ with a diffeological vector space complement, $\mathfrak{K}$, satisfying*

---

*i)* $\chi-1(C^\infty(I, \mathfrak{H})) \subseteq C_0^\infty(I, H)$ ;

*ii)* given $h \in H$ there exists a smooth path $f : [0,1] \mapsto H$ such that $f(0)=e$, $f(1)=h$, and $\chi(f) \in C^\infty(I, \mathfrak{H})$;

*iii)* given any $k \in \mathfrak{G}/\mathfrak{H}$, there exists a smooth Lie algebra homomorpism $\phi$, of $\mathfrak{G}/\mathfrak{H}$ into a the Lie algebra $\mathfrak{S}$ of a regular diffeological Lie group $S$ such that $\phi(k) \neq 0$.

Then $G/H$ is a Lie group with Lie algebra $\mathfrak{G}/\mathfrak{H}$.

**Lemma 1.1.** *Under the hypotheses of the previous theorem, given $u(t), v(t) \in C_0^\infty(I, G)$ suppose that $\chi(u) - \chi(v) \in \mathfrak{H}$. Then there exists $w(t) \in C_0^\infty(I, H)$ such that $v(t) = u(t)\dot{w}(t)$.*

This lemma implies

**Theorem 1.6.** *Under the hypotheses of theorem ??, H and G/H are regular diffeological Lie groups.*

## 2 On the integrability of some generalized Kac-Moody algebras

Inspired by Borcherds we define a generalized Kac-Moody algebra by the given of a

1) A real Hilbert space $\{H, <;>\}$ together with a continuous non degenerate symmetric bilinear form $(\ldots,\ldots)$.

2) Suppose given a countable set of elements $h_i \in H$ such that $(h_i, h_j) \leq 0$ if $i \neq j$ and such that if $(h_i, h_i)$ is positive, then $2(h_i, h_j)/(h_i, h_i)$ is an integer;

3) $A = ((h_i, h_j))$ is called a generalized symmetrized Cartan matrix of real numbers.

The generalized Kac-Moody algebra $\mathfrak{G} = \{\mathfrak{G}(A), H\}$ is the Lie algebra generated by the vector space H and symbols $e_i$ and $f_i$, with defining relations:

$[H, H] = 0$;

$[e_i, f_j] = \delta_{ij} h_i$;

$[h, e_i] = (h_i, h)e_i, [h, f_i] = -(h_i, h)f_i$;

If $a_{ii} > 0$, then $(ad(e_i)^{1-2a_{ij}/a_{ii}}(e_j)) = 0 = (ad(f_i)^{1-2a_{ij}/a_{ii}}(f_j))$.

We have the root space decomposition of $\mathfrak{G}(A) = \sum_{\alpha \in \mathfrak{H}^*} \mathfrak{G}_\alpha$, where

$$\mathfrak{G}_\alpha = \{x \in \mathfrak{G}(A) | [h, x] = \alpha(h)x, \forall h \in \mathfrak{H}\}.$$

Let's recall that the height of a root is defined by

$$ht(\sum_{i=1}^n z_i \alpha_i) = \sum_{i=1}^n |z_i|.$$

The non-degenerate symmetric bilinear form $(\ldots,\ldots)$ on H extends uniquely to a non-degenerate invariant symmetric bilinear form on $\mathfrak{G}(A)$ called the <u>canonical</u> form. Now define $\omega : \mathfrak{G}(A) \mapsto \mathfrak{G}(A)$ by $\omega(e_i) = -f_i, \omega(f_i) = -e_i, \omega(h) = -h, h \in H$ and set $(x,y)_0 = -(x, \omega(y))$.

We now define on $\mathfrak{G}(A)$ the following positive definite inner product

$$(\alpha_+ + h + \alpha_-, \beta_+ + k + \beta_-)_1 = (\alpha_+, \beta_+)_0 + <h,k> + (\alpha_-, \beta_-)_0.$$

We shall suppose that $\sum_i |a_{ij}|^2 < E < \infty, \forall j$.

We have [5] $\exists\, C > 0$ so that

$$(\star)||[z, z']||_1 \le C(||d(z)||_1 ||z'||_1 + ||z||_1 ||d(z')||_1)$$

where $d$ is a derivation of $\mathcal{G}(A)$ satisfying $d(x) = ht(\alpha)x$ for $x \in \mathcal{G}_\alpha(A)$.

Now consider the subspace $\bar{\mathfrak{G}}(A) \subset \prod \mathfrak{G}_\alpha(A)$, where $\alpha$ ranges over all the roots of $\mathfrak{G}(A)$, characterized by $\{M_\alpha\} \in \bar{\mathfrak{G}}(A)$ if and only if given any $t \ge 1$ there exists a constant $K_t$ so that

$$\sum_{\alpha, ht(\alpha)=n} ||M_a|| \le K_t e^{-t|n|}.$$

On the vector space $\bar{\mathfrak{G}}(A)$, consider the Hausdorff locally convex topology generated by the following fundamental system of neighborhoods of $0 \in \bar{\mathfrak{G}}(A)$: to each pair of positive real numbers $(t, k)$, let

$$U(t,k) = \{\{M_\alpha\} \in \bar{\mathfrak{G}}(A) : \sum_{\alpha, ht(\alpha)=n} ||m_\alpha|| < k e^{-t|n|}\}$$

With this topology $\bar{\mathfrak{G}}(A)$ is a Fréchet space. In this topology, the bounded sets are characterized by

**Lemma 2.1.** *$B \subset \bar{\mathcal{G}}(A)$ is bounded if and only if for any positive real number $t$ there exists a constant $B_t$ such that $\{M_\alpha\} \in B$ implies $|| \sum_{\alpha, ht(\alpha)=n} M_\alpha || \le B_t e^{-tn}$.*

It is not difficult to verify that all bounded sets are relatively compact. In [14] we prove

**Proposition 2.1.** *Given $\{g_\alpha\}, \{h_\alpha\} \in \bar{\mathcal{G}}(A)$ then $[\{g_\alpha\}, \{h_\alpha\}] = \{k_\alpha\} \in \bar{\mathfrak{G}}(A)$; further, $[\{g_\alpha\}, \{h_\alpha\}]$ determines a topological Lie algebra structure on $\bar{\mathfrak{G}}(A)$.*

We recall that a Hausdorff, sequentially complete, locally convex topological vector space V is called strongly bornological (resp. bornological) when any subset (resp. convex subset) absorbing all the bounded subsets of V is a neighborhood of the origin. In general in this paper the topological vector spaces with which we deal will be strongly bornological. Note that metrizable locally convex topological vector spaces are strongly bornological as are countable inductive limits of strongly bornological spaces.

We shall make use of a sequence of functions defined iteratively as follows for $t > 0$ and $q \in Z$:

$$F_0(t, q) = \sum_{n \in Z} (|n + q| + |n|)e^{-t(|n+q|+|n|)},$$

$$F_{k+1}(t, q) = \sum_n (|n + q| + |n|)F_k(t, n + q)e^{-t|n|}.$$

One readily verifies the following lemmas and corollaries.

**Lemma 2.2.** $F_0(t, q)$ is bounded by some constant $D$ for $1 \leq t_0 < s < t, q \in Z$ and satisfies $F_0(t, q) \leq F_0(t - s, q) \cdot e^{-s|q|}$.

**Corollary 2.1.** $F_k(t, q) \leq F_k(t - s, q) \cdot e^{-s|q|}$.

**Lemma 2.3.** There exists a constant $\kappa \geq 1$ such that

$$\sum_{n \in Z} (|n + q| + |n|)e^{-t(|n|)} < \kappa |q|$$

for all $t > t_0 \geq 1$, where $\kappa$ is independent of $t$.

**Corollary 2.2.** $F_n(t, q) \leq D(\kappa |q|)^n$.

**Lemma 2.4.** Define $\phi(t, q) = \sum_{n \geq 0} \frac{F_n(t,q)}{n!}$, then $|\phi(t, q)| \leq De^{\kappa |q|}$.

**Lemma 2.5.** Given $\xi = \{\xi_\alpha\}, \eta = \{\eta_\alpha\} \in \bar{\mathcal{G}}(A)$ such that $\|\xi_n\| \leq X_t e^{-tn}$ and $\|\eta_n\| \leq Y_t e^{-t|n|}$, where $\xi_n = \sum_{\alpha, ht(\alpha)=n} \xi_\alpha, \eta_n = \sum_{\alpha, ht(\alpha)=n} \eta_\alpha$. Then $\|[\xi, \eta]\| \leq CX_t Y_t F_0(t, q)$, where $C$ is the constant of $\{\star\}$.

**Lemma 2.6.** Given the hypotheses of lemma 2.5, then

$$\|[[\xi, \eta_1], \eta_2]\| \leq C^2 X_t Y_t^2 F_1(t, q),$$

where $\eta_1, \eta_2$ satisfy the $\star$ of lemma 2.5.

By iteration, we obtain

**Lemma 2.7.** Given the hypotheses of lemma 2.6, then

$$\|(ad_{\eta_k} \circ \cdots \circ ad_{\eta_1}(\xi))\| \leq C^k X_t Y_t^k F_{k-1}(t, q),$$

where $\eta_k, \circ \ldots, \eta_1$, and $\xi$ satisfy lemma 2.5.

We have [13]

**Theorem 2.1.** Let $A$ be a graded Hausdorff complete locally convex topological Lie algebra $A$ with a strongly bornological underlying vector space topology such that given any bounded set $B \subset A$ there exists a balanced convex bounded set $C$ such that $\sum_{n=k}^{\infty} (1/n!)B^n \to 0$ as $k \to \infty$ in $A_C = \bigcup_n nC$ with the $C$-gauge norm topology; that is, $|\alpha|_C = \inf\{\lambda > 0 : \alpha \in \lambda C\}$, where $B^{n+1} = [B, B^n]$. Then the canonical diffeological structure on $A$ is pre-integrable.

Lemmas 2.2-2.7 imply

**Theorem 2.2.** *If $\mathcal{G}(A)$ is a generalized Kac-Moody algebra, then $\bar{\mathcal{G}}(A)$ is an integral pre-integrable diffeological Lie algebra.*

**Definition 2.1.** Given two smooth paths $f, g : I \mapsto L$ into a diffeological Lie algebra, $L, f, g : I \mapsto L$ we say then are Lie homotopic when there exists smooth maps from the square $V, W : I \times I \mapsto L$ such that $V(t, 0) = f, V(t, 1) = g, W(0, s) \equiv 0, W(1, s) \equiv 0, V_s - W_t = [V, W]$.

Now let $E \subset H$, where H is a Hilbert space be an Hausdorff, sequentially complete, locally convex topological vector space with a topology finer than the induced topology from the Hilbert space, we suppose E furnished with the canonical diffeology from its locally convex topology, and let End(E) be the algebra of smooth linear endomorphisms of E. We suppose that End(E) has the diffeology induced from F(E, E)

One verifies [14] that

**Proposition 2.2.** *End(E) is a diffeological algebra.*

and

**Lemma 2.8.** *Given $a > 0$, let $p : (-a, a) \mapsto End(E)$ be a path through the identity, $p(0) = id \in End(E)$; if $[p] = 0 \in T_e(End(E))$, then $D_{t=0}p(t)(e) = 0$ for each $e \in E$.*

*Remark 2.1.* Lemma 2.5 implies that there exists a canonical homomorphism $\kappa : T_e(\text{End(E)}) \mapsto \text{End(E)}$.

In this section we shall give conditions which imply that a representation $\mathcal{L} \mapsto \text{End(E)}$, of the Lie algebra $\mathcal{L}$ of a simply connected regular diffeological Lie group L, is the derivative of a multiplicative homomorphism of $L \mapsto \text{End(E)}$, where End(E) has the canonical Lie algebra structure: $[A, B] = AB - BA$.

Notation: Given a diffeological vector space $E$ and a smooth map $f : I \mapsto E$, where $I = [0, 1]$, designate by $\hat{f}$ the convex hull of $f(I) \subset E$.

**Definition 2.2.** Let $E \subset H$, where H is a Hilbert space, be an Hausdorff, sequentially complete, bornological, locally convex topological vector space with a topology rendering continuous the canonical injection into H, and suppose that L is a regular diffeological Lie group with diffeological Lie algebra, $\mathcal{L}$. A smooth representation $F : \mathcal{L} \mapsto \text{End(E)}$, where End(E) has the diffeology induced by the function space diffeology, is called regular when given any smooth map, $f : I \mapsto \mathcal{L}$ and any bounded subset, $B \subset E$, we have that

$$\sum_{n=k}^{\infty} (1/n!) L(\hat{f})^n(B) \to 0 \text{ as } k \to \infty$$

in $A_C = \bigcup_n nC$ with the C-gauge norm topology; that is, $|\alpha|_C = inf\{\lambda > 0 : \alpha \in \lambda C\}$, where C is some bounded convex subset of E.

We are able to define the fundamental group of a diffeological space by means of the connected components, given by smooth arcs, of the diffeological

space $\Omega(X, x_0) \equiv \{f \in F(R, X) : f(t) = x_0, t < \epsilon, t > 1 - \epsilon, 0 < \epsilon < 1/2\}$ .
By iteration one defines the higher homotopy groups and can show that an exact sequence of homotopy groups for diffelogical fibrations exists. Iglesias has shown that for a connected diffeological space a universal covering space exists; that is, he establishes an unique principal fibration with discrete fiber isomorphic to the fundamental group. By a straightforward generalization of classical strategies one can show that a connected diffeological group has an unique universal covering diffeological group such that the covering map is a smooth homomorphism.

As a direct consequence of the theory of linear differential equations on bornological spaces [13] we have

**Lemma 2.9.** *Let L be a simply connected regular diffeological Lie group, $H : \mathfrak{L} \mapsto End(E)$ a smooth regular representation of Lie algebras, and $f : I \mapsto \mathfrak{L}$ a smooth function, then there exists an unique smooth path $k : I \to End(E)$ such that $k'(t) = H(f(t))(k(t))$, such that $k(0) = id \in End(E)$.*

We have [14]

**Theorem 2.3.** *Let L be a regular simply connected diffeological Lie group, under the hypotheses of lemma 6 there exists an unique multiplicative smooth homomorphism $\phi : L \mapsto End(E)$ such that $\kappa \circ T_e(\phi) = H$, where $\kappa : T_e(End(E)) \mapsto End(E)$ is the homomorphism of remark ??.*

# References

1. Borcherds, R., Generalized Kac-Moody Lie Algebras, J. of Algebra 115, 501-512 (1988)
2. Bourbaki, N., Espaces Vectoriels Topologiques, Chap. III and IV.
3. Dazord, P., Lie Groups and Algebras in Infinite Dimensions: a new approach. XXXIIIrd Taniguchi Symposium, Symplectic Geometry and its applications, 1993
4. Donato,P. in Géométrie Symplectique & Mécanique, Springer 14 16 ( ), 84-104
5. Garland, H., The Arithmetic Theory of Loop Groups, Publ. Math. I.H.E.S. 52 (1980), 5-136.
6. Goodman, R. and Wallach, N. R., Structure and unitary cocyle representation of loop groups and the group of diffeomorphisms of the circle, Jour. r. angew. Math. 347 (1984), 69-133.
7. Iglesias, P., Thesis, 1985
8. Kac, V. and Peterson, D. H., Unitary structure in representations of infinite dimensional groups and a convexity theorem, invert. math. 76 (1984), 1-14.

9. Kac, V., Infinite Dimensional Lie Algebras, Cambridge Univ. Press (1985).

10. Kac, V., "Constructing Groups Associated to Infinite-Dimensional Lie Algebras, Infinite Dimensional Groups with Applications," edited by V. Kac, Springer-Verlag, 1985.

11. Leslie, J., On the Lie Subgroups of Infinite Dimensional Lie Groups, Bull. of AMS Jan. (1987)

12. Leslie, J. On a Super Lie Group Structure for the Group of $G^\infty$ Diffeomorphisms of a compact $G^\infty$ supermanifold Geometry & Physics, 1997

13. Leslie, J., Lie's Third Theorem in Infinite Dimensions, Algebras, Groups And Geometries 14, 359-405 (1997).

14. On A Diffeological Group Realization of Certain Generalized Symmetrizable Kac-Moody Lie Algebras (to appear)

15. Mathieu, 0., Construction du groupe de Kac-Moody et applications, CR Acad. Sci. série 1 (1988), 227-230.

16. Milnor, J., Remarks on Infinite Dimensional Lie Groups, Proc. of Summer School on Quantum Gravity (1983).

17. Moody, R. V. and Teo, K. L., Tits' systems with crystallographic Weyl groups, J. Algebra 21 (1972), 178-190.

18. Omori, H., A remark on non-enlargible Lie algebras, J. Math. Soc. Japan 33 (1981), 707-710.

19. Pressley, A. and Segal, G., Loop Groups, Oxford Science Publications (1986).

20. Ray, Urmie, Generalized Kac-Moody Algebras and Some Related Topics, BAMS, VOL. 38, No. 1, Jan. 2001

21. Sikorski, R., Differential Modules, Colloquium Mathematicum, vol. XXIV, 1971

22. Souriau, J. M., in "Feuilletages et Quantification geometrique". Travaux en Cours, Hermann,1984, 365-398

23. Suto, K., Differentiable Vectors and Analytic Vectors in Completions of Certain Representation Spaces of a Kac-Moody Algebra,, Proc. Japan Acad. 63 Ser. A (1987).

24. Tits, J., Groups and Group Functions Attached to Kac-Moody Data, Lecture Notes in Mathematics no. 1111 (1984), 193-223.

# THE LIE GROUP OF FOURIER INTEGRAL OPERATORS ON OPEN MANIFOLDS

RUDOLF SCHMID

*Department of Mathematics,*
*Emory University*
*Atlanta, GA 30322, USA*
*E-mail: rudolf@mathcs.emory.edu*

We give a survey on the Lie group structures of pseudodifferential operators and Fourier integral operators together with their associated diffeomorphism groups. Our goal is to describe the case of open manifolds, but we first review the compact case and then outline the difficulties that arise for non-compact manifolds and show how we solved them.

## Introduction

The theory of pseudodifferential operators and Fourier integral operators on compact manifolds is well established and their applications in mathematical physics well known, see for example Hörmander [8], Duistermaat [5], Treves [12]. For open (non compact) manifolds this is not the case, and that's what I would like to focus on in this paper. We are interested in the geometry of the spaces of pseudodifferential operators and Fourier integral operators and their associated diffeomorphism groups. Unfortunately, the theory on open manifolds is very technical and much more complicated than in the compact case. I will give here only an overview of the results and refer for technicalities to the published papers Eichhorn and Schmid [6,7]. Instead of going through the details of the construction of the Lie group structures of pseudodifferential and Fourier integral operators on open manifolds I will first review (part A) the compact case and then explain (part B) what goes wrong in the non compact case and how we fixed these problems.

## A-COMPACT CASE

In 1985 we proved the following theorem Adams, Ratiu , Schmid [1,2,3]:

**Theorem 0.1.** *The group $FIO_*(M)$ of invertible Fourier integral operators on a* **compact** *manifold $M$ is a graded $\infty$-dim Lie group with graded $\infty$-dim Lie algebra $\Psi DO(M)$ of pseudodifferential operators on $M$. $FIO_*(M)$ is an $\infty$-dim principal fiber bundle over the base manifold $Diff_\theta(\dot{T}^*M)$ of contact transformations of $\dot{T}^*M$ with gauge group $\Psi DO_*(M)$ of invertible* *pseudodifferential operators.*

$$\Psi DO_*(M) \longrightarrow FIO_*(M)$$

$$\downarrow$$

$$Diff_\theta(\dot{T}^*M)$$

In 2002 we proved an analogous theorem for **open** manifolds, see Eichhorn and Schmid [7]. I'll first explain the construction of this principal fiber bundle and these Lie groups in the **compact** case.

# 1 The Lie group structure of $FIO_*(M)$ on compact manifolds

## 1.1 What are Fourier integral operators ?

Let me call Fourier integral operators $FIO$ for short.
• $FIO$ generalize pseudodifferential operators (let me denote them by $\Psi DO$) which in turn are a generalization of differential operators (call them $DO$). So we have the following inclusions as sets

$$FIO \supset \Psi DO \supset DO$$

Let me start to explain the space $\Psi DO$.
Let $M$ be a compact manifold. $P$ is a classical $\Psi DO$ of order $m$ on $M$ ($P \in \Psi DO_m$) if it is locally of the following form : for any $u \in C_c^\infty(M)$

$$Pu(x) = (2\pi)^{-n} \int \int e^{i(x-y)\cdot\xi} p(x,\xi)u(y)dy\ d\xi \tag{1}$$

where $p(x,\xi)$ ia a *classical* symbol of order $m$, i.e. it is a smooth function having an asymptotic expansion

$$p(x,\xi) \sim \sum_{j=m}^{-\infty} p_j(x,\xi)\ , \tag{2}$$

where each term $p_j(x,\xi)$ is homogeneous of degree $j$ in $\xi$, i.e $p_j(x,t\xi) = t^j p_j(x,\xi)$ , $t > 0$. The leading term $a_m(x,\xi)$ is called the *principal symbol* of the operator $P$ and it is globally defined on $T^*M$. For more general symbol classes see eg. Hörmander [8]. We restrict ourselves to classical $\Psi DO$s.
• Special case: If the symbol is a polynomial in $\xi$ i.e. of the form

$$p(x,\xi) = \sum_{|\alpha|\leq m} p_\alpha(x)\xi^\alpha \tag{3}$$

then $P$ is a differential operator of order $m$, so $\Psi DO \supset DO$ .

$\Psi DO$s are oscillatory integrals and highly singular but they are nice operators in the following sense:

(i) They are invariant under diffeomorphisms, which implies that they are defined on the whole manifold $M$ as bounded linear operators $P : C^\infty(M) \to C^\infty(M)$,

(ii) they are pseudolocal i.e they preserve the singular support *sing supp* $(Pu) \subset$ *sing supp* $(u)$, and they preserve the wave front sets $WF$, $WF(Pu) \subset WF(u)$,

(iii) they extend as bounded linear operators to Sobolev spaces $P : H_c^s(M) \to H_c^{s-m}(M)$,

(iv) they are closed under composition and the order is additive , i.e. if $P \in \Psi DO_m$ and $Q \in \Psi DO_n$ then $P \circ Q \in \Psi DO_{m+n}$ ,

(v) the space of all pseudodifferential operators of all orders $\Psi DO = \bigcup_m \Psi DO_m$ is an $\infty$-dim graded Lie algebra, i.e. if $P \in \Psi DO_m$ and $Q \in \Psi DO_n$ then the commutator $[P, Q] = P \circ Q - Q \circ P \in \Psi DO_{m+n-1}$.

Note: The space $\Psi DO_1$ of all $\Psi DO$s of order one is itself an $\infty$-dim Lie algebra.

*Remark* 1.1. The principal symbol of the commutator $[P, Q]$ is given by $\{p_m, q_n\}$ the canonical Poisson bracket on $T^*M$ of the principal symbols $p_m$ of $P$ and $q_n$ of $Q$. This property leads to a quantization theory via $\Psi DO$s !

Since in infinite dimensions, not every Lie algebra has a corresponding Lie group, a natural question is the following:

**QUESTION**: Are there LIE GROUPS corresponding to these $\infty$-dim Lie algebras $\Psi DO$ and $\Psi DO_1$ ? i.e. are there $\infty$-dim Lie groups $\mathcal{G}$ , $\mathcal{G}_1$ which have $\Psi DO$ resp. $\Psi DO_1$ as their Lie algebras ?

**Warning**: The classical Lie theorems do not hold in $\infty$ dimensions ! i.e. not every $\infty$-dim Lie algebra automatically has a corresponding Lie group !

Nevertheless the answer is :

**ANSWER**: YES ! $\mathcal{G} = FIO_*$ the group of invertible Fourier integral operators and and $\mathcal{G}_1 = (FIO_o)_*$ the group of invertible Fourier integral operators of order zero are the $\infty$-dim. Lie groups with Lie algebras $\Psi DO$ and $\Psi DO_1$ respectively.

Let me now explain what $FIO$s are:

$A$ is a (classical) Fourier integral operator of order $m$ $(A \in FIO_m)$ if it is locally of the following form : for any $u \in C_c^\infty(M)$

$$Au(x) = (2\pi)^{-n} \int \int e^{i\varphi(x,y,\xi)} a(x,\xi) u(y) dy \, d\xi \qquad (4)$$

where $a(x,\xi)$ is classical symbol of order $m$ as in (2) i.e. $a(x,\xi) \sim \sum_{j=m}^{-\infty} a_j(x,\xi)$ and $\varphi(x,y,\xi)$ is a nondegenerate phase function. We don't give the precise definition but one of the basic properties of the phase function $\varphi(x,y,\xi)$ is that it is homogeneous of degree one in $\xi$ and that it generates a conic Lagrangian submanifold $\Lambda$ in $T^*(M \times M) - 0$. We restrict ourselves to those $FIOs$ for which this Lagrangian submanifold $\Lambda$ is the graph of a diffeomorphism $\eta : \dot{T}^*M \to \dot{T}^*M$, (where $\dot{T}^*M = T^*M - 0$, minus the zero section), i.e. $\Lambda = graph(\eta)$ . This means that the phase function $\varphi$ is locally a generating function of a homogeneous canonical transformation $\eta : \dot{T}^*M \to \dot{T}^*M : \eta^*\omega = \omega, \eta(t\alpha) = t\eta(\alpha)$ which is equivalent that $\eta$ preserves $\theta$, the canonical 1-form on $T^*M$, i.e. $\eta^*\theta = \theta$. These diffeomorphisms are called *quantomorphisms*.

- Special case: If $\varphi(x,y,\xi) = (x - y) \cdot \xi$ then $\Lambda$ is the diagonal in $\dot{T}^*M \times \dot{T}^*M$ and $\varphi$ generates $\eta = id : \dot{T}^*M \to \dot{T}^*M$ . So the $FIO$ $A$ with this pace function $\varphi$ is a $\Psi DO$. That implies that $FIO \supset \Psi DO \supset DO$.

Again like $\Psi DOs$ the $FIOs$ are nice operators in the following sense:

(i) They are invariant under diffeomorphisms, which implies that they are defined on the whole manifold $M$ as bounded linear operators $A : C^\infty(M) \to C^\infty(M)$,

(ii) $A : C^\infty(M) \to C^\infty(M)$ extends as bounded linear operator to Sobolev spaces $A : H_c^s(M) \to H_c^{s-m}(M)$,

(iii) they move the wave front sets (the complement of where $u$ is $C^\infty$) by a canonical relation $\Lambda$, i.e. $WF(Au) \subset \Lambda \circ WF(u)$, where $\Lambda$ is the conic Lagrangian submanifold $\Lambda \subset \dot{T}^*M \times \dot{T}^*M$ locally generated by the phase function $\varphi(x,y,\xi)$ of $A$,

(iv) they are closed under composition and the order is additive i.e. if $A_1 \in FIO_m$ and $A_2 \in FIO_n$ then $A_1 \circ A_2 \in FIO_{m+n}$. Moreover if $\eta_1$ is generated by the phase function $\varphi_1$ of $A_1$ and $\eta_2$ is generated by the phase function $\varphi_2$ of $A_2$ then $\eta_1 \circ \eta_2$ is generated by the phase function of $A_1 \circ A_2$. If $A \in FIO_m$ is invertible with canonical transformation $\eta$ then $\eta^{-1}$ is generated by the phase function of $A^{-1} \in FIO_{-m}$.

**Example 1.1.** Let $f : M \to M$ be a diffeomorphism. Then the pullback

$$f^*u(x) = (2\pi)^{-n} \int \int e^{i(f(x)-y)\cdot\xi} u(y) dy \ d\xi \tag{5}$$

defines a $FIO$ $f^* : C^\infty(M) \to C^\infty(M)$ whose phase function generates the canonical cotangent lift $T^*f : \dot{T}^*M \to \dot{T}^*M$.

### 1.2 The exact sequence of groups

Let's denote $FIO$ the space of all Fourier integral operators of all orders , i.e. $FIO = \bigcup_m FIO_m$ and $FIO_*$ the set of all invertible $FIO$s. Then $FIO_*$ is a group with group operation being composition of $FIO$s. We denote by $\Psi DO_*$ the group of invertible $\Psi DO$s and by $Diff_\theta^\infty = \{\eta \in Diff^\infty(\dot{T}^*M) \mid \eta^*\theta = \theta\}$ the diffeomorphism group of quantomorphisms.

We have an exact sequence of groups

$$I \to \Psi DO_* \xrightarrow{j} FIO_* \xrightarrow{p} Diff_\theta^\infty \to id \tag{6}$$

where the first map $j$ is just the inclusion, the second map $p$ is defined as follows: for $A \in FIO_*$ , $p(A) = \eta \in Diff_\theta^\infty$ such that $\eta$ is generated by the phase function $\varphi$ of $A$ . This map $p$ is surjective since we restrict ourselves to those $FIO$s with canonical relations $\Lambda = graph(\eta)$, for some $\eta \in Diff_\theta^\infty$. The kernel of $p$ equals $ker \ p = \Psi DO_*$ , since the phase function $\varphi(x, y, \xi) = (x - y) \cdot \xi$ generates the identity map $id : \dot{T}^*M \to \dot{T}^*M$. Hence the sequence (6) is an exact sequence of groups. We want to make this into an exact sequence of LIE GROUPS !

Since the order of $FIO$s and $\Psi DO$s is additive under composition (group operation) the invertible operators of order zero $(\Psi DO_0)_*$ and $(FIO_0)_*$ also form groups. Let's consider the corresponding exact sequence of zero order groups and their "Lie algebras"

$$Groups \qquad I \to (\Psi DO_0)_* \xrightarrow{j} (FIO_0)_* \xrightarrow{p} Diff_\theta^\infty \to id \tag{7}$$

$$Lie \ algebras \qquad 0 \longrightarrow \Psi DO_0 \hookrightarrow \Psi DO_1 \longrightarrow Vec_\theta^\infty \to 0 \tag{8}$$

We don't know the Lie groups structures of these groups yet, but we can give some heuristic arguments why these Lie algebras in (8) are the corresponding "Lie algebras" of the "Lie groups" in (7). The formal Lie algebras are identified with the corresponding tangent spaces at the identity. As we have for finite dimensional Lie groups $G$, the Lie algebra $\mathbf{g} \simeq T_e G$, $e \in G$ the identity. We have

1) Since the space $\Psi DO_0$ of all $\Psi DO$s of order zero is a vector space and the group $(\Psi DO_0)_*$ of invertible elements is an open set in $\Psi DO_0$, its tangent space at $I \in (\Psi DO_0)_*$ is isomorphic to $\Psi DO_0$, i.e $T_I(\Psi DO_0)_* = \Psi DO_0$.

2) For the "Lie algebra" of $(FIO_0)_*$, consider a curve in $(FIO_0)_*$ through the identity $I$. Taking its derivative (at $I$) we obtain a $\Psi DO$ of order one, since the phase functions of the $FIO$s are homogeneous of degree one, i.e by the chain rule we're adding to the symbol of an $FIO$ of order zero a term of order one and evaluating at $I$, which gives a $\Psi DO$ of order one.

3) For $Diff_\theta^\infty$, the infinitesimal condition of $\eta^*\theta = \theta$ is $L_X\theta = 0$, i.e. the Lie algebra of $Diff_\theta^\infty$ is $Vec_\theta^\infty = \{X \; vector \; field \mid L_X\theta = 0\}$.

*Remark* 1.2. Note that we also have the corresponding exponential maps form the Lie algebras into the groups, but in infinite dimensions they don't locally generate the Lie groups (unlike in finite dim) i.e we cannot obtain our desired Lie group structures by exponentiating the corresponding Lie algebras

$$(\Psi DO_0)_* \neq exp(\Psi DO_0) \;, \quad (FIO_0)_* \neq exp(\Psi DO_1) \;, \quad Diff_\theta^\infty \neq exp(Vec_\theta^\infty).$$

## 1.3 The principal fiber bundle

We have to *construct* these Lie groups and the idea is the following: Note that a $\Psi DO$ is determined by its symbol and these form a linear infinite dimensional vector space, whereas a $FIO$ is characterized by a symbol and a phase function; the symbols again form a linear space but the phase functions are associated with diffeomorphisms which form an infinite dimensional (non linear) manifold. So the idea is to construct an $\infty$-dimensional principal fiber bundle such that

- base manifold $= Diff_\theta^\infty(\dot{T}^*M)$
- total space $= (FIO_0)_*$
- fiber $= p^{-1}(\eta) \simeq (\Psi DO_0)_* =$ gauge group

This construction is done in 7 steps (see Adams, Ratiu, Schmid [1,2] and Ratiu, Schmid [10])

- Step 1: We show that $Diff_\theta^\infty = \lim_{\infty \leftarrow s} Diff_\theta^s$ is an ILH Lie group.
- Step 2: We show that $(\Psi DO_0)_* = \lim_{\infty \leftarrow s} (\Psi DO_0^s)_*$ is an ILH Lie group .
- Step 3: We construct a local section of the exact sequence (7), which gives the local product structure at the identity.
- Step 4: We check compatibility conditions to obtain $(FIO_0)_*$ as a topolog-

ical group.

• Step 5: We check smoothness of the chart transition maps to obtain $(FIO_0)_*$ as a smooth manifold.

• Step 6: We check smoothness of group multiplication and inversion to obtain $(FIO_0)_*$ as a ILH Lie group.

• Step 7: We carry the Lie group structure of $(FIO_0)_*$ to the whole space $FIO_*$.

### 1.4 Step 1: $Diff_\theta^\infty$ as ILH Lie group

We show that $Diff_\theta^\infty = \varprojlim_{\infty \leftarrow s} Diff_\theta^s$ is the inverse limit of Hilbert Lie groups, called ILH Lie group. On each group $Diff_\theta^s$ of Sobolev class $H^s$ diffeomorphisms we construct a Hilbert Lie group structure and take the inverse limit as $s \to \infty$ to obtain $Diff_\theta^\infty = \varprojlim_{\infty \leftarrow s} Diff_\theta^s$ as ILH Lie group, (Ratiu , Schmid [10]). There is one complication here. In order to show that $Diff_\theta^s$ is a submanifold of $Diff^s(\dot{T}^*M)$ we need to show that $Diff^s(\dot{T}^*M)$ is a manifold in the first place, but $\dot{T}^*M$ is never compact, even if $M$ is. So we go to the cosphere bundle $ST^*M \subset T^*M$, which is compact and carries an induced contact form $\theta_S$. Each quantomorphism $\eta : \dot{T}^*M \to \dot{T}^*M$ with $\eta^*\theta = \theta$ is in one to one correspondence with an induced contact transformation, i.e. a pair $(\varphi, h)$ , where $h : ST^*M \to \mathbf{R}$ is a smooth function and $\varphi : ST^*M \to ST^*M$ is a diffeomorphism with $\varphi^*\theta_S = h\theta_S$ . We showed in Ratiu , Schmid [10] that $Diff_\theta^s$ is isomorphic to the semidirect product

$$Diff_\theta^s(ST^*M) := \{(\varphi, h) \in Diff^s(ST^*M) \rtimes C^s(ST^*M) \mid \varphi^*\theta_S = h\theta_S\}.$$

The composition of quantomorphisms goes into the semidirect product of contact transformation as follows: Let $\eta_1 \mapsto (\varphi_1, h_1)$ and $\eta_2 \mapsto (\varphi_2, h_2)$, then $(\eta_1 \circ \eta_2) \mapsto (\varphi_1, h_1) \cdot (\varphi_2, h_2) = (\varphi_1 \circ \varphi_2, h_2(h_1 \circ \varphi_2))$.

With this we show that $Diff_\theta^\infty = \varprojlim_{\infty \leftarrow s} Diff_\theta^s$ is an ILH Lie group.

### 1.5 Step 2: $(\Psi DO_0)_*$ as ILH Lie group

We show that $(\Psi DO_0)_* = \varprojlim_{\infty \leftarrow s} (\Psi DO_0^s)_*$ is an ILH Lie group . The topology on $\Psi DO$ is determined by the symbols, which are smooth functions and form a vector space but if we want to define a topology on $\Psi DO_0$ directly, we would end up with a Frechet space because each $P \in \Psi DO_0$ has a symbol of the form $p(x, \xi) = \sum_{j=0}^{-\infty} p_j(x, \xi)$, so we would have to have control over an infinite number of functions and their derivatives, and an infinite product of Hilbert spaces is no longer a Hilbert space. So what we do is to cut the symbol at the term $p_{-k}$ for some fixed $k < \infty$. In terms of operators we look at the quotient spaces $\Psi DO_{m,k} = \Psi DO_m / \Psi DO_{-k-1}$. Similarly for $FIO$ we take $FIO_{m,k}(\eta) = FIO_m(\eta) / FIO_{-k-1}(\eta)$ and $FIO_{m,k} = \bigcup_\eta FIO_{m,k}(\eta)$ , where $FIO_m(\eta) = \{ A \in FIO_m \mid p(A) = \eta \in Diff_\theta^\infty \}$. Composition is still well defined in $\Psi DO_{0,k}$ and $FIO_{0,k}$ and we denote by $(\Psi DO_{0,k})_*$ and $(FIO_{0,k})_*$ the groups of invertible elements in $\Psi DO_{0,k}$ and $FIO_{0,k}$ respectively. We still have the exact sequence of groups:

$$I \to (\Psi DO_{0,k})_* \overset{j}{\hookrightarrow} (FIO_{0,k})_* \overset{p}{\to} Diff_\theta^\infty \to id \qquad (9)$$

For $P \in \Psi DO_{0,k}$ with symbol $p(x,\xi) = p_m(x,\xi) + \cdots + p_{-k}(x,\xi)$ we define the norm by

$$\|P\|_{m+k,s}^2 = \|\tilde{p}_m\|_{s+k+m}^2 + \|\tilde{p}_{m-1}\|_{s+k+m-1}^2 + \cdots + \|\tilde{p}_k\|_s^2 \qquad (10)$$

where $\tilde{p}_{m-j}$ is the restriction of $p_{m-j}$ to the cosphere bundle $ST^*M$ and $\|\tilde{p}_{m-j}\|_{s+k+m-j}^2$ is the $H^{s+k+m-j}$-Sobolev norm on $ST^*M$. Let $\Psi DO_{m,k}^s$ denote the completion of $\Psi DO_{m,k}$ with respect of this norm and for $m = 0$ we denote by $(\Psi DO_{0,k}^s)_*$ the group of invertible elements in $\Psi DO_{0,k}^s$.

**Theorem 1.1 (Adams, Ratiu , Schmid [1]).** *For each $s > n$ the group $(\Psi DO_{0,k}^s)_*$ is a Hilbert Lie group with Lie algebra $\Psi DO_{0,k}^s$. That means $(\Psi DO_{0,k}^s)_*$ is a smooth ($C^\infty$) Hilbert manifold with smooth group operations. Moreover the inverse limit $(\Psi DO_{0,k})_* = \lim_{\infty \leftarrow s} (\Psi DO_{0,k}^s)_*$ is an ILH Lie group.*

At the end of the day we will take the limit $k \to \infty$ !

### 1.6 Step 3: The local section

Now we have Lie group structures on both ends of our exact sequence (7) and we can piece these together via a local section $\sigma$

$$\sigma : U \subset Diff_\theta^\infty \to (FIO_0)_* \qquad (11)$$

where $U$ is a neighborhood of $id \in Diff_\theta^\infty$. Since $FIOs$ are defined locally, this requires to construct a global writing of $FIOs$ which are associated to diffeomorphisms $\eta \in Diff_\theta^\infty$ which are close to the identity. This is quite elaborate and done in Adams, Ratiu, Schmid [2]. There we define the local section $\sigma$ as follows: Let $\eta \in Diff_\theta^\infty$ close to the identity and define $\sigma(\eta)$ by

$$\sigma(\eta)u(x) := (2\pi)^{-n} \int_{T_x^*M} \int_{B_\delta(x)} \chi(x,y) e^{i\varphi_H(\alpha_x,y)} u(y) |\det \exp_*| dy d\xi \qquad (12)$$

where $\varphi_H(\alpha_x, y) = \varphi_0(\alpha_x, y) + H(\alpha_x)$, with $\varphi_0(\alpha_x, y) = \alpha_x \cdot \exp_x^{-1}(y)$ is the global generating function of the identity $id \in Diff_\theta^\infty$ and $H$ a Sobolev small perturbation. Explicitly $\varphi_H(\alpha_x, y) = \alpha_x \cdot \exp_x^{-1}(y) - \alpha_x \cdot \exp_x^{-1}(p(\eta^{-1}(\alpha_x)))$. $B_\delta(x)$ is the neighborhood where $\exp_x$ is a diffeomorphism and $\chi$ a bump function with $supp\ \chi \subset B_\delta(x)$.

The operator $\sigma(\eta)$ is a $FIO$ with smooth phase function $\varphi_H$ and amplitude $a = 1$. Moreover, $\sigma(\eta)$ is invertible modulo smoothing operators since $\eta$ is invertible and its principal symbol is $a = 1$, hence $\sigma(\eta) \in (FIO_0)_*$. Furthermore, $p\sigma(\eta) = \eta$, hence $\sigma$ is a local section of the exact sequence (7).

Now we can use this local section $\sigma$ to give $(FIO_0)_*$ the local product structure

$$p^{-1}(U) \simeq \sigma(U) \times (\Psi DO_0)_* .$$

We define the topology around the identity in $(FIO_{0,k})_*$ by the bijection $\Phi : p^{-1}(U^{2t}) \to U^{2t} \times (\Psi DO_{0,k}^{2(t-k)})_*$ , $\Phi(A) = (p(A), A \circ \sigma(p(A))^{-1})$ and $\Phi^{-1}(\eta, P) = P \circ \sigma(\eta)$, where $U^{2t} = U \cap Diff_\theta^{2t}$.

This defines a local chart at the identity $I \in (FIO_0)_*$.


## 1.7 Step 4: $(FIO_0)_*$ as topological group

To define the topology on $(FIO_{0,k})_*$ we move the open sets $p^{-1}(U^{2t})$ by right translations. Complete this topological space in the right-uniform structure and denote it by $(FIO_{0,k}^t)_*$. For each $t > n/2$ we obtain $(FIO_{0,k}^t)_*$ as a topological group and $(FIO_{0,k})_* = \bigcap_t (FIO_{0,k}^t)_*$ with the inverse limit topology is a topological group as well. To prove this, we have to show that the map $(A, B) \mapsto AB^{-1}$ is continuous for any $A, B \in (FIO_{0,k}^t)_*$. This amounts to show that the following map in local coordinates is continuous:

$$(U^{2t} \times \Psi DO_{0,k}^{2(t-k)}) \times (U^{2t} \times \Psi DO_{0,k}^{2(t-k)}) \to (U^{2t} \times \Psi DO_{0,k}^{2(t-k)})$$

$$((\eta_1, P_1), (\eta_2, P_2)) \mapsto (\eta_1 \circ \eta_2^{-1}, P_1 \sigma(\eta_1) \sigma(\eta_2)^{-1} P_2^{-1} \sigma(\eta_1 \circ \eta_2^{-1})^{-1}) \quad (13)$$

which involves a careful study of products of $FIOs$.


## 1.8 Step 5: $(FIO_{0,k})_*$ as smooth manifold

Overlap conditions in local charts give conditions on $\sigma$ to make $(FIO_0)_*$ into a smooth manifold. To prove smoothness of the transition maps between local charts we have to show that the following map is differentiable

$$(U^{2t} \cdot \alpha \cap U^{2t} \cdot \beta) \times (\Psi DO_{0,k}^{2(t-k)})_* \to (\Psi DO_{0,k}^{2(t-k)})_*$$

$$(\eta, P) \mapsto P\sigma(\eta \circ \alpha^{-1})AB^{-1}\sigma(\eta \circ \beta^{-1})^{-1} \tag{14}$$

for any $A, B \in (FIO_{0,k}^t)_*$, where $\alpha = p(A), \beta = p(B)$. The symbol calculus shows that this map is of class $C^t$, hence $(FIO_{0,k}^t)_*$ is a smooth manifold of class $C^t$.

## 1.9 Step 6: $(FIO_0)_*$ as ILH Lie group

We check that the group multiplication and inversion

$$\mu : (FIO_0)_* \times (FIO_0)_* \to (FIO_0)_* \quad , \quad \mu(A, B) = A \circ B \tag{15}$$

$$\nu : (FIO_0)_* \to (FIO_0)_* \quad , \quad \nu(A) = A^{-1} \tag{16}$$

are differentiable which makes $(FIO_0)_*$ into a Lie group.
To show that group multiplication in $(FIO_{0,k}^t)_*$ is smooth we have to show that the following map is differentiable

$$((U^{2(t+s)} \cdot \alpha) \times (\Psi DO_{0,k}^{2(t+s+k)})_*) \times ((U^{2(t+s)} \cdot \beta) \times (\Psi DO_{0,k}^{2(t+s+k)})_*) \to$$

$$((U^{2(t+s)} \cdot (r\alpha \cdot \beta)) \times (\Psi DO_{0,k}^{2(t+s+k)})_*)$$

$$((\eta_1, P_1), (\eta_2, P_2)) \mapsto$$

$$(\eta_1 \circ \eta_2, P_1\sigma(\eta_1 \circ \alpha^{-1})AP_2\sigma(\eta_2\beta^{-1})A^{-1}\sigma(\eta_1\eta_2\beta^{-1}\alpha^{-1})^{-1}) \tag{17}$$

for any $A \in (FIO_{0,k}^{t+s})_*, B \in (FIO_{0,k}^t)_*$, where $\alpha = p(A), \beta = p(B)$.

This makes $(FIO_0)_* = \lim_{\infty \leftarrow s} (FIO_0^s)_*$ into an ILH Lie group.
The situation is similar to diffeomorphism groups. $(FIO_0)_*$ is an ILH Lie group and right multiplication is smooth but left multiplication and inversion are only continuous. We state the precise smoothness estimates below.

*1.10   Step 7: FIO$_*$ as Lie group*

Now we have the zero order operators $(FIO_0)_*$ as a Lie group . In order to obtain a Lie group structure on all $FIO_*$ we use the Laplace operator to identify $(1 - \Delta)^{m/2} : (FIO_0)_* \xrightarrow{\sim} (FIO_m)_*$. Multiplication is smooth between the appropriate spaces. Piecing this together for all $m$ makes the group of invertible Fourier integral operators $FIO_* = \cup_m(FIO_m)_*$ a graded direct limit of ILH Lie groups with Lie algebra $\Psi DO$ the space of pseudodifferential operators.

The final result for the compact case now is the following:

**Main Theorem 1 (M. Adams, T. Ratiu, R. Schmid [1,2]).** *The group $FIO_*(M)$ of invertible Fourier integral operators on a* **compact** *manifold $M$ is a graded $\infty$-dim ILH-Lie group with graded $\infty$-dim Lie algebra $\Psi DO(M)$ of pseudodifferential operators on $M$. $FIO_*(M)$ is an $\infty$-dim principal fiber bundle over the base manifold $Diff_\theta^s(\dot{T}^*M)$ of contact transformations of $\dot{T}^*M$ with gauge group $\Psi DO_*(M)$ of invertible pseudodifferential operators.*

More precisely,

**Theorem 1.2 (M. Adams, T. Ratiu, R. Schmid [1,2]).** *The smoothness estimates are as follows:*

- **(A)** The Lie group structure: $(FIO_{0,k})_*$ viewed as the projective limit $\lim_{\infty \leftarrow t}(FIO_{0,k}^t)_*$ for $t$ strictly larger than $\max(2n, n + 2(k + 1))$ is an ILH Lie group. Explicitly:

  (i) For each $t > \max(2n, n + 2(k + 1))$, $(FIO_{0,k}^t)_*$ is a $C^t$ manifold modeled on $H^{t+1}(\dot{T}^*M) \times \Psi DO_{0,k}^{t-2(k+1)}$.

  (ii) $(FIO_{0,k}^t)_*$ is a topological group and $(FIO_{0,k})_* = \lim_{\infty \leftarrow t}(FIO_{0,k}^t)_*$ is a topological group with the inverse limit topology.

  (iii) The inclusions $j : (FIO_{0,k}^{t+1})_* \hookrightarrow (FIO_{0,k}^t)_*$ are $C^t$.

  (iv) Group multiplication $\mu : (FIO_{0,k})_* \times (FIO_{0,k})_* \rightarrow (FIO_{0,k})_*$ , $\mu(A, B) = A \circ B$, extends $C^q$, $q = \min(t, p)$ to $\mu : (FIO_{0,k}^{t+p})_* \times (FIO_{0,k}^t)_* \rightarrow (FIO_{0,k}^t)_*$.

  (v) Inversion $\nu : (FIO_{0,k})_* \rightarrow (FIO_{0,k})_*$ , $\nu(A) = A^{-1}$, extends $C^q$, $q = \min(t, p)$ to $\nu : (FIO_{0,k}^{t+p})_* \rightarrow (FIO_{0,k}^t)_*$.

*(vi) For $A \in (FIO_{0,k}^t)_*$, right multiplication $R_A : (FIO_{0,k}^t)_* \rightarrow (FIO_{0,k}^t)_*$, $R_A(B) = B \circ A$ is $C^t$.*

- (B) The Lie algebra structure: *The Lie algebra of the ILH Lie group $(FIO_{0,k})_*$ is the ILH Lie algebra $(\Psi\tilde{D}O_{1,k})_* = \lim_{\infty \leftarrow t} (\Psi\tilde{D}O_{1,k}^t)_*$ with $t > \max(2n, n+2(k+1))$, where $\Psi\tilde{D}O_{1,k}^t = \{ P \in \Psi D O_{1,k}^t$ with pure imaginary principal symbol $\}$. Explicitly:*

  *(i) For each $t > \max(2n, n + 2(k + 1))$, $\Psi\tilde{D}O_{1,k}^t$ is a Hilbert space.*

  *(ii) $\Psi\tilde{D}O_{1,k} = \lim_{\infty \leftarrow t} \Psi\tilde{D}O_{1,k}^t$ is a Frechet space with the inverse limit topology.*

  *(iii) The inclusions $j : \Psi\tilde{D}O_{1,k}^{t+1} \rightarrow \Psi\tilde{D}O_{1,k}^t$ are continuous and dense.*

  *(iv) The Lie bracket (commutator) is bilinear, continuous, antisymmetric*

  $$[\,,\,] : \Psi\tilde{D}O_{1,k}^{s+2} \times \Psi\tilde{D}O_{1,k}^{t+2} \rightarrow \Psi\tilde{D}O_{1,k}^{\min(s,t)} \quad : \quad [P,Q] = PQ - QP$$

  *for all $t > \max(2n, n+2(k+1))$ and satisfies the Jacobi identity on $\Psi\tilde{D}O_{1,k}^{\min(s,t,r)}$ for elements in $\Psi\tilde{D}O_{1,k}^{s+4} \times \Psi\tilde{D}O_{1,k}^{t+4} \times \Psi\tilde{D}O_{1,k}^{r+4}$.*

## B-NON-COMPACT CASE

*There is exactly one thing that works in the non-compact case like for compact manifolds: nothing!* (my collaborator J. Eichhorn)

In this second part I will illustrate what goes wrong for non compact manifolds and how we fixed it, beginning with diffeomorphism groups of open manifolds, then discussing $\Psi DO$ and $FIO$ for non compact manifold. Details can be found in Eichhorn and Schmid [6,7].

## 2  Diffeomorphisms of NON-COMPACT manifolds

**Example of what goes wrong:** Let $M$ and $N$ be compact manifolds. Then a map $f : M^m \rightarrow N^n$ is of Sobolev class $H^s$ if and only if the local representatives $f_j^i : U_i \subset \mathbf{R}^m \to V_j \subset \mathbf{R}^n$ are of class $H^s$, where $M = \bigcup(U_i, \phi_i)$, $N = \bigcup(V_j, \psi_j)$, $f_j^i := \psi_j \circ f \circ \phi_i^{-1}$. These covers are **finite** if $M, N$ are compact. By Sobolev's imbedding theorem, this definition is invariant if $s > \frac{m}{2} + 1$. In the compact case we can define the distance between $f, g \in H^s(M)$ by

$$d^s(f, g) := \left( \sum_{i,j} \|f_j^i - g_j^i\|_s^2 \right)^{\frac{1}{2}}.$$

These definitions are meaningless if $M, N$ are open !

Our solution to these type of problems is to replace compactness by **bounded geometry**.

## 2.1 Bounded geometry

### Idea of bounded geometry
- We have control over the metric and its derivatives up to any order.
- We have control over the mappings and their derivatives by the metric, i.e. the diffeomorphisms are adapted to the bounded geometry.

**Definition 2.1.** A Riemannian manifold $(M^n, g)$ has *bounded geometry of order* $k$, $0 \leq k \leq \infty$, if $M$ has a positive injectivity radius $r_{inj}(M)$ and the curvature tensor $R$ and all is derivatives up to order $k$ are uniformly bounded; i.e the following two conditions $(I)$ and $(B_k)$ are satisfied:

$$(I): \ r_{inj}(M) = \inf_{x \in M} r_{inj}(x) > 0$$

$$(B_k): \ |\nabla^i R| \leq C_i \ , \ 0 \leq i \leq k.$$

*Remark* 2.1. Equivalent conditions are:
$(I) \Leftrightarrow$ there exists a ball around 0 in $\mathbf{R}^n$ which is domain of normal (geodesic) coordinates **for all** $x \in M$.
$(B_k) \Leftrightarrow$ there exists a constant $d_k$ (independent of $x \in M$) such that $\|g_{ij}\|_{C^k} \leq d_k$ in any normal coordinate system
$\Leftrightarrow \|\Gamma_{ij}^m\|_{C^{k-1}} \leq d_k$ in any normal coordinate system.
**Example 2.1.** Examples of manifolds with bounded geometry are compact manifolds, Lie groups, homogeneous spaces, covering spaces of Riemannian manifolds, leaves of foliations of compact manifolds.

Given any open manifold $M^n$ and $k \geq 0$, then there exists a complete Riemannian metric $g$ on $M^n$ satisfying the conditions $(I)$ and $(B_k)$; i.e there is **no** topological obstruction for a metric with bounded geometry of any order.

We now adapt all our previous constructions to the bounded geometry of the underlying manifolds. All the manifolds are assumed to have bounded geometry, i.e. satisfy conditions $(I)$ and $(B_k)$.

## 2.2   Bounded maps $C^{\infty,r}(M,N)$

Consider $(M,g),(N,h)$ open, complete Riemannian manifolds satisfying $(I)$ and $(B_k)$ and $f \in C^\infty(M,N)$. Assume $r \leq k$. We denote by $C^{\infty,r}(M,N)$ the set of all $f \in C^\infty(M,N)$ satisfying

$$|df|_r := \sum_{i=0}^{r-1} \sup_{x \in M} |\nabla^i df|_x < \infty. \tag{18}$$

Equivalently: $f \in C^{\infty,r}(M,N) \Leftrightarrow \frac{\partial^\alpha}{\partial x^\alpha} f^\nu$ is uniformly bounded in any normal coordinate system; $|\alpha| \leq r, 1 \leq r \leq k$.

We define the topology (uniform structure) on $C^{\infty,r}(M,N)$ as follows. Two bounded maps $f,g \in C^{\infty,r}(M,N)$ are said to be close iff there exists a vector field $\xi$ along $f$ (i.e. $\xi \in C^\infty(f^*TN)$) with small $H^p$-Sobolev norm $\|\xi\|_{p,r} < \varepsilon$, with $0 < \varepsilon < \frac{1}{2}r_{inj}(N)$, such that $g(x) = exp_{f(x)}\xi(x)$; where

$$\|\xi\|_{p,r} := \left( \int \sum_{i=0}^{r} |\nabla^i t|_x^p dvol_x(g) \right)^{\frac{1}{p}}. \tag{19}$$

The completion of $C^{\infty,r}(M,N)$ with respect to this uniform structure we denote by $C^{p,r}(M,N)$.

**Theorem 2.1 (Eichhorn, Schmid [6]).** *Let* $1 < p < \infty$ *,$r \leq k$, $r > \frac{n}{p}+1$. Then the completion $C^{p,r}(M,N)$ is a $C^{k+1-r}$-Banach manifold, and for $p = 2$ it is a Hilbert manifold.*

## 2.3   The bounded diffeomorphism group $Diff^{p,r}(M)$ .

**Problem:** We have an additional problem, namely $C^{p,r}(M) \bigcap Diff(M)$ is **not** a group. If $f$ is a bounded diffeomorphism then $f^{-1}$ need not to be bounded, i.e. $f \in C^{p,r}(M) \bigcap Diff(M) \not\Rightarrow f^{-1} \in C^{p,r}(M)$ . We need an additional assumption which implies that the bounded diffeomorphisms $Diff^{p,r}(M)$ are open in $C^{p,r}(M,M)$ and each component is a $C^{k+1-r}$-Banach

58

manifold, and for $p = 2$ it is a Hilbert manifold. We assume that $|\lambda|_{min}(df)$, the absolute value of the eigenvalues of the Jacobian of $f$, is bounded away from 0, i.e.

$$|\lambda|_{min}(df) > 0. \tag{20}$$

Set

$$Diff^{p,r}(M) := \{f \in C^{p,r}(M,M) \mid f \text{ is bijective, and } |\lambda|_{min}(df) > 0\}. \tag{21}$$

**Theorem 2.2 (J. Eichhorn, R. Schmid [6]).** *Let $(M^n, g)$ be an open, oriented, complete Riemannian manifold satisfying $(I)$, $(B_k)$ and let $r > \frac{n}{p} + 1$. Then*
*a) $Diff^{p,r}(M)$ is open in $C^{p,r}(M,M)$, in particular each component is a $C^{k+1-r}$-Banach manifold, and for $p = 2$ it is a Hilbert manifold.*
*b) If $(M,g)$ satisfies $(B_\infty)$ then $Diff^{p,\infty}(M) = \lim_{\leftarrow} Diff^{p,r}(M)$ is an ILB - Lie group; and for $p = 2$ it is an ILH - Lie group.*

Next we discuss some important subgroups of the diffeomorphism group, the volume preserving- , symplectic- and contact diffeomorphism groups.

## 2.4   Volume preserving and symplectic diffeomorphisms

Assume $(M^n, g)$ is an open manifold satisfying the conditions $(I)$ and $(B_k), k \geq m \geq r > \frac{n}{2} + 1$. Let $\omega$ be a $C^m$-bounded closed non-degenerate $q$-form with $\inf_{x \in M} |\omega|_x^2 > 0$, and consider the group of form preserving bounded diffeomorphisms $Diff^{p,r}_\omega(M) = \{f \in Diff^{p,r}(M) \mid f^*\omega = \omega)\}$. The technique to show that $Diff^{p,r}_\omega(M)$ is a closed submanifold of $Diff^{p,r}(M)$ is to show that the map

$$\psi : Diff^{p,r}(M) \to \{exact\ q - forms\ \} : \psi(f) := f^*(\omega) \tag{22}$$

is a submersion, and then argue that $Diff^{p,r}_\omega(M) = \psi^{-1}(\omega)$ is a close submanifold. For this one needs a Sard's theorem and a Hodge decomposition theorem which allows one to conclude that the exterior derivative operator is closed. This is not automatically given for open manifolds, so we need an additional condition. Let $\Delta_q$ be the Laplace operator acting on $q$-forms, $\sigma_e(\Delta_q)$ its essential spectrum and $\sigma_e(\Delta_q|_{(ker\Delta_q)^\perp})$ the essential spectrum of $\Delta_q$ restricted to the orthogonal complement of its kernel. We assume that its $g.l.b.$ $\inf \sigma_e(\Delta_1|_{(ker\Delta_1)^\perp})$ is bounded away from 0; i.e. we assume the following the spectral condition

$$\inf \sigma_e(\Delta_1|_{(ker\Delta_1)^\perp}) > 0. \tag{23}$$

Then under this spectral condition (23) , the group of form preserving diffeomorphisms $Diff_\omega^{p,r}(M)$ is a $C^{k-r+1}$ submanifold of $Diff^{p,r}(M)$, see Eichhorn, Schmid [6].

We are now in position to state the following theorem; the technique of the proof of which is similar to the compact case, but technically more complicated.

**Theorem 2.3 (Eichhorn, Schmid [6]).** *Under the above assumptions (20) and (23) we have:*
*a) $Diff_\omega^{p,\infty} = \lim_{\leftarrow r} Diff_\omega^{p,r}$ is an ILH-Lie group with Lie algebra consisting of divergence free ($q = n$), or locally Hamiltonian ($q = 2$) vector fields $\xi$ with finite Sobolev norm $\|\xi\|_{p,r}$ for all $r$.*
*b) $Diff_\omega^{p,r}$ is an infinite dimensional Riemannian manifold, with (weak) metric*

$$g(X,Y)_{id} = \int_M (X,Y)_x dvol_x(g). \qquad (24)$$

## 2.5 Contact transformations of the restricted cotangent bundle $\dot{T}^*M$

We will use the techniques developed in the compact case, so we restrict the contact transformations of $\dot{T}^*M$ to the cosphere bundle $S(T^*M) \subset \dot{T}^*M$. But now $S(T^*M)$ is not compact but carries an induced metric of bounded geometry. If $(M^n, g_M)$ is an open, oriented, complete Riemannian manifold satisfying $(I)$, $(B_k)$ then the Sasaki metric on $T^*M$ satisfies $(I)$, $(B_{k-1})$. Let $\pi : T^*M \to M$ be the projection and $K$ the connection map of the Levi-Civita connection in $T^*M$. Then the Sasaki metric $g_{T^*M}$ on $T^*M$ is defined by

$$g_{T^*M}(X,Y) = g_M(\pi_* X, \pi_* Y) + g_M(KX, KY), \quad X,Y \in TT^*M.$$

Then $g_S := g_{T^*M}|S(T^*M)$ , the restriction of the Sasaki metric to the cosphere bundle $S(T^*M) \subset \dot{T}^*M$, satisfies $(I)$, $(B_{k-1})$. Let $\theta$ be the canonical 1-form on $T^*M$ and consider

$$Diff_\theta^{p,r} = \{f \in Diff^{p,r}(\dot{T}^*M)|\ f^*\theta = \theta\ \}.$$

Like in the compact case we showed in Eichhorn, Schmid [7] that $Diff_\theta^{p,s}$ is isomorphic to the semidirect product

$$Diff_\theta^{p,s}(S(T^*M)) :=$$

$$\{(\eta_S, h) \in Diff^{p,s}(S(T^*M)) \rtimes C^{p,s}(S(T^*M)) \mid \eta_S^*\theta_S = h\theta_S\}$$

We proved the following

**Theorem 2.4 (Eichhorn, Schmid [7]).**

$$Diff_\theta^{p,\infty} = \lim_{\infty \leftarrow r} Diff_\theta^{p,r}$$

*is an ILH-Lie group.*
*Remark 2.2.* This is the space of phase functions for the Fourier integral operators!

## 3 Pseudodifferential operators and Fourier integral operators on open manifolds

Pseudodifferential operators, $\Psi DOs$, and Fourier integral operators, $FIOs$, are well defined for any manifold, open or closed. But on open manifolds the spaces of these operators don't have any reasonable structures. Moreover, many theorems for $\Psi DOs$ or $FIOs$ on closed manifolds become wrong or don't make any sense in the open case, e.g. certain mapping properties between Sobolev spaces of functions are wrong. The situation rapidly changes if we restrict ourselves to bounded geometry and adapt these operators to the bounded geometry. This means, roughly speaking, that the family of local symbols together with their derivatives should be uniformly bounded. For $FIOs$ we additionally restrict ourselves to comparatively smooth Lagrangian submanifolds $\Lambda$ of $\dot{T}^*M \times \dot{T}^*M$ and phase functions also adapted to the bounded geometry. Then the corresponding spaces $\Psi DO$ and $FIO$ have similar properties as in the compact case and we can use the same ideas as before to construct Lie group structures. This is technically quite complicated and we refer for details to Eichhorn and Schmid [7] and give here only the general ideas. Fourier integral operators are characterized by their symbols $a(x,\xi)$ and their phase functions $\varphi(x,y,\xi)$

$$FIO \quad Au(x) = (2\pi)^{-n} \int \int e^{i\varphi(x,y,\xi)} a(x,\xi) u(y) dy \, d\xi \; .$$

The main idea is to require the following for symbols and phase functions:
- **Symbols:** The family of local symbols $a(x,\xi)$ together with their derivatives should be uniformly bounded, i.e. they should be elements of $C^{p,r}(U \times \mathbf{R}^n, \mathbf{R})$.
- **Phase functions:** The phase functions $\varphi(x,y,\xi)$ should locally generate canonical transformations in the space $Diff_\theta^{p,r}(\dot{T}^*M)$.

## 3.1 Uniform pseudodifferential operators $\mathcal{U}\Psi DO(M)$

More precisely we define a class of uniformly bounded symbols as follows:
Let $B = B_\epsilon(0) \subset \mathbf{R}^n$ be the ball which is the domain of normal (geodesic) coordinates for **all** $m \in M$. This exists thanks to our assumption (I) of bounded geometry ! Let $q \in \mathbf{R}$ and define symbol families $\{a_m\}_{m \in M}$ with $a_m \in C^\infty(B \times \mathbf{R}^n)$, such that

$$|\partial_\xi^\alpha \partial_x^\beta a_m(x, \xi)| \le C_{\alpha,\beta}(1 + |\xi|)^{q-|\alpha|} \tag{25}$$

where $C_{\alpha,\beta}$ is independent of $m \in M$. In addition $a_m(x, \xi)$ is classical if $a_m(x, \xi) \sim \sum_{j=0}^\infty a_{m,q-j}(x, \xi)$.

Each symbol family $\{a_m\}_{m \in M}$ defines a family of operators $a_m(x, D_x)$ : $C_c^\infty(B) \to C^\infty(B)$,

$$a_m(x, D_x)u(x) := (2\pi)^{-n} \int_{\mathbf{R}^n} \int_B a_m(x, \xi)u(y)dyd\xi. \tag{26}$$

We define the space $\mathcal{U}\Psi DO_q(M)$ of **uniform** pseudodifferential operators of order $q$ as follows:

$A \in \Psi DO_q(M)$ with Schwartz kernel $\mathcal{K}_A$ belongs to $\mathcal{U}\Psi DO_q(M)$ iff
(i) ex. const. $C_A > 0$, s.t. $\mathcal{K}_A(x, y) = 0$ for $d(x, y) > C_A$
(ii) $\mathcal{K}_A$ is smooth outside the diagonal of $M \times M$
(iii) for any $\delta > 0, i, j$ ex.const $C_{\delta,i,j} > 0$ s.t.

$$|\nabla_x^i \nabla_y^j \mathcal{K}_A(x, y)| \le C_{\delta,i,j} , \quad d(x, y) > \delta$$

(iv) define $A_m := exp_m^* \circ A \circ (exp_m^*)^{-1} : C_c^\infty(B) \to C^\infty(B)$ has the form $A_m = a_m(x, D_x) + R_m$. ($R_m$ smoothing operator). Note that $-k \le q \Rightarrow \mathcal{U}\Psi DO_{-k-1} \subset \mathcal{U}\Psi DO_q$ and denote $\mathcal{U}\Psi DO_{q,k} := \mathcal{U}\Psi DO_q/\mathcal{U}\Psi DO_{-k-1}$ and $\mathcal{U}\Psi DO_{q,k}^s$ its completion in the $H^s$ Sobolev norm.

## 3.2 Uniform Fourier integral operators $\mathcal{U}FIO(M)$

Now we define a **uniform** family of nondegenerate phase functions $\{\varphi_m\}_{m \in M}$ by requiring that each $\varphi_m(x, y, \xi)$ is locally generating a canonical transformation $\eta_m : \dot{T}^*M \to \dot{T}^*M$ : $\eta_m^*\theta = \theta$, with $\eta_m \in Diff_\theta^{p,r}(\dot{T}^*M)$. With this we define the class of **uniform** Fourier integral operators of order $q$, denoted by $\mathcal{U}FIO_q$ as follows: $A \in \mathcal{U}FIO_q$ iff $A : C_c^\infty(M) \to C^\infty(M)$ and locally $A_m := exp_m^* \circ A \circ (exp_m^*)^{-1} : C_c^\infty(B) \to C^\infty(B)$ is of the form

$$A_m u(x) = (2\pi)^{-n} \int_{\mathbf{R}^n} \int_B e^{i\varphi_m(x,y,\xi)} a_m(x, y, \xi)u(y)dyd\xi \tag{27}$$

where $a_m$ is a uniformly bounded classical symbol of order $q$ and $\varphi_m$ a uniform phase function. Let $\mathcal{U}FIO_{q,k} := \mathcal{U}FIO_q/\mathcal{U}FIO_{-k-1}$.

We get the exact sequence of groups

$$I \to (\mathcal{U}\Psi DO_{0,k})_* \xrightarrow{j} (\mathcal{U}FIO_{0,k})_* \xrightarrow{p} Diff_\theta^{p,r} \to id \qquad (28)$$

## 3.3  7 steps to the Lie group $\mathcal{U}FIO$

Now we follow the same ideas as in the compact case: Performing Steps 1,2...7 as before to construct ILH Lie groups structures on these spaces.

- Step 1: We show that $Diff_\theta^{p,\infty} = \lim\limits_{\infty \leftarrow s} Diff_\theta^{p,s}$ is an ILH Lie group.

- Step 2: We show that $(\mathcal{U}\Psi DO_0)_* = \lim\limits_{\infty \leftarrow s} (\mathcal{U}\Psi DO_0^s)_*$ is an ILH Lie group .

- Step 3: We construct a local section of the exact sequence (28) to obtain the local product structure around the identity.

- Step 4: We check compatibility conditions to obtain $(\mathcal{U}FIO_0)_*$ as a topological group.

- Step 5: We check smoothness of the chart transitions maps to obtain $(\mathcal{U}FIO_0)_*$ as a smooth manifold.

- Step 6: We check smoothness of group multiplication and inversion to obtain $(\mathcal{U}FIO_0)_*$ as a ILH Lie group.

- Step 7: We carry the Lie group structure of $(\mathcal{U}FIO_0)_*$ to the whole space $\mathcal{U}FIO_*$.

The openness of the underlying manifold always requires additional considerations and additional estimates which are quite technical and different from the compact case. In addition, the ILH-Lie algebras are quite different from the ones in the compact case. We refer to Eichhorn, Schmid [7] for details and state the main result:

**Main Theorem 2 (Eichhorn, Schmid [7]).** *Assume $(M^n, g)$ is an open Riemannian manifold satisfying the conditions $(I)$ and $(B_\infty)$ of bounded geometry and the condition*

$$\inf \sigma_e(\triangle_1(S(T^*M)), g_S|_{(ker\triangle_1)^\perp}) > 0$$

*Then for any $k \in \mathbf{Z}_+$*

*1. $Diff_{\theta,0}^{2,\infty}(\dot{T}^*M) = \lim\limits_{\leftarrow s} Diff_{\theta,0}^{2,r}(\dot{T}^*M)$ , $r \geq n+1$, is an ILH Lie group.*

*2. $(\mathcal{U}\Psi DO_{0,k})_* = \lim\limits_{\leftarrow s}(\mathcal{U}\Psi DO_{0,k}^s)_*$ , $s \geq n+1$, is an ILH Lie group and each $(\mathcal{U}\Psi DO_{0,k}^s)_*$ is a smooth Hilbert Lie group.*

*3. $(\mathcal{U}FIO_{0,k})_* = \lim\limits_{\leftarrow t}(\mathcal{U}FIO_{0,k}^t)_*$ , $t > \max\{2n, n+2(k+1)\}\}$ is an ILH Lie* group with the following properties.

a. $(\mathcal{U}FIO_{0,k}^t)_*$ is a $C^t$ Hilbert manifold modeled on $C^{2,t+1}(S(T^*M)) \times (\mathcal{U}\Psi DO_{0,k}^{t-2(k+1)})_*$.

Each component of $(\mathcal{U}FIO_{0,k}^t)_*$ is modeled by $C^{2,t+1}(S) \oplus C^{2,k+t-(2k+1)}(S) \oplus \cdots \oplus C^{2,t-2(k+1)}(S)$.

b. The inclusion $(\mathcal{U}FIO_{0,k}^{t+1})_* \hookrightarrow (\mathcal{U}FIO_{0,k}^t)_*$ is a $C^t$ map

c. The group multiplication $\mu$ is a $C^l$ map, $l = \min\{r, t\}$

$$\mu : (\mathcal{U}FIO_{0,k}^{t+r})_* \times (\mathcal{U}FIO_{0,k}^t)_* \longrightarrow (\mathcal{U}FIO_{0,k}^t)_* \,, \quad \mu(A, B) = A \circ B$$

d. The inversion $\nu$ is a $C^l$ map, $l = \min\{r, t\}$

$$\nu : (\mathcal{U}FIO_{0,k}^{t+r})_* \longrightarrow (\mathcal{U}FIO_{0,k}^t)_* \,, \quad \nu(A) = A^{-1}$$

e. Right multiplication $R_A$ by $A \in (\mathcal{U}FIO_{0,k}^t)_*$ is a $C^t$ map

$$R_A : (\mathcal{U}FIO_{0,k}^t)_* \longrightarrow (\mathcal{U}FIO_{0,k}^t)_* \,, \quad R_A(B) = B \circ A.$$

Remark 3.1. The Lie algebra of the ILH Lie group $(\mathcal{U}FIO_{0,k})_*$ is the ILH Lie algebra $(\tilde{\mathcal{U}}\Psi DO_{1,k})_* = \lim_{\infty \leftarrow t} (\tilde{\mathcal{U}}\Psi DO_{1,k}^t)_*$ with commutator properties like in the compact case.

## References

1. M.R. Adams, T Ratiu and R. Schmid, *Math. Ann.* **273**, 529 (1986).
2. M.R. Adams, T Ratiu and R. Schmid, *Math. Ann.* **276**, 19 (1986).
3. M.R. Adams, T Ratiu and R. Schmid, in *MSRI Publications,* 4,ed. V. Kac (1985).
4. A. Banyaga, *The Structure of Classical Diffeomorphism Groups* (Kluwer Acad. Publ., Dordrecht, 1997).
5. J.J. Duistermaat, *Fourier Integral Operators* (Birkhäuser , Boston, 1966).
6. J. Eichhorn and R. Schmid, *Ann. Global Analysis and Geometry* **14**, 147 (1996).
7. J. Eichhorn and R. Schmid, *Comm. in Analysis and Geometry,* vol **9**, no 5 (2001), 983–1040.
8. L. Hörmander, *The Analysis of Linear Partial Differential Operators* I-IV, (Springer Verlag, Berlin, 1990).
9. H. Omori, *Infinite-Dimensional Lie Groups* (AMS , Translations of Math. Monographs, **158**, Providence, 1997).

10. T.Ratiu and R. Schmid, *Math. Z.* **177**, 81 (1981).

11. R. Schmid , *Infinite Dimensional Hamiltonian Systems* (Biblionopolis, Napoli,1987).

12. F. Treves, *Introduction to pseudodifferential and Fourier integral operators I, II*, (Plenum Press, New York, NY., 1980).

# ON SOME PROPERTIES OF LEIBNIZ ALGEBROIDS

AISSA WADE

*Department of Mathematics, Penn State University*
*University Park, PA 16802, USA.*
*e-mail: wade@math.psu.edu*

We give a characterization of Leibniz algebroids in terms of coboundary operators. Also, we show that, for any Leibniz algebroid $E$ with base $M$, there is a class in the first Leibniz cohomology $HL^1(E)$, which is intrinsically associated with the Leibniz algebroid structure on $E$ and agrees with the modular class of a Lie algebroid.

## 1  Introduction

The notion of a Lie algebroid can be viewed as a generalization of the tangent bundle of a smooth manifold. In recent years, a number of extremely interesting examples of Lie algebroids have been found in differential geometry as well as in physics. Recall that a *Lie algebroid* is a vector bundle $E \to M$ together with a bracket on sections of $E$, and a bundle map $\varrho : E \to TM$, called the anchor, whose extension to sections of these vector bundles is a homomorphism of Lie algebras such that $[X, fY] = f[X, Y] + (\varrho(X)f)Y$.

Leibniz algebroids are weakened versions of Lie algebroids, where the bilinear operation on sections of the considered vector bundle is not necessarily skew-symmetric, and $\varrho$ is just a homomorphism of Leibniz algebras. The concept of a Leibniz algebroid was introduced by Ibáñez, León, Marrero and Padrón (see [3]) in their study of Nambu-Poisson manifolds. A basic non-trivial example is the vector bundle $E = TM \times \mathbb{R}$ with the $\mathbb{R}$-bilinear operation $[(X_1, f_1),\ (X_2, f_2)] = ([X_1, X_2], X_1 f_2)$ and $\varrho(X, f) = X$. This construction can be generalized to any Lie algebroid. In other words, every Lie algebroid corresponds to a Leibniz algebroid.

The aim of this paper is to study the main properties of Leibniz algebroids. The crucial tool developed here is the analogue of the classical Cartan differential calculus for Leibniz algebroids, which is essentially based on Loday's definition of Leibniz cohomology (see [7]). We show in Theorem 3.1 that there is a one-to-one correspondence between the Leibniz algebroid structures on a vector bundle $E \to M$ and some coboundary operators $\delta$ on $C^\bullet(\Gamma(E), C^\infty(M))$. Theorem 3.1 generalizes a result of [9].

It is well-known that any Lie algebroid structure on a vector bundle

$E \to M$ gives rise to a graded Lie bracket (called algebraic Schouten bracket) on the exterior algebra $\Gamma(\Lambda^\bullet E)$ generated by sections of $E$. This Schouten bracket together with the exterior multiplication define a Gerstenhaber algebra structure on $\Gamma(\Lambda^\bullet E)$ (see [10]). It is of interest to explore whether such results can be extended to the context of Leibniz algebroids. It turns out that this phenomenon is no longer true for general Leibniz algebroids. Precisely, we show that there is a "generalized" Schouten bracket $[\![\cdot,\cdot]\!]$ on $\Gamma(\Lambda^\bullet E)$. That is, a non-skew-symmetric version of the algebraic Schouten bracket for which $[\![X,\cdot]\!]$ is a derivation when $X \in \Gamma(E)$, but not for some $X \in \oplus_{n \geq 2}\Gamma(\Lambda^n E)$.

We found an amazing aspect of the theory of Leibniz algebroids. Namely, as in the classical case of Lie algebroids, using the interior product and the Lie derivative operator, one can define the modular class of a Leibniz algebroid.

In this paper, Section 2 is preparatory, giving a back-ground material for Leibniz algebroids. Section 3 forms the main part of the work.

## 2 Graded Leibniz Algebras and Cohomology

In this section, we recall some known facts on Leibniz algebras.

A *graded Leibniz algebra* is a graded vector space $\mathcal{A} = \oplus_{i \in \mathbb{Z}}\mathcal{A}_i$ over a field $\mathbf{k}$ of characteristic zero with a bilinear operation $[\ ,\ ] : \mathcal{A} \times \mathcal{A} \to \mathcal{A}$ of degree $\ell$ such that

$$[a,[b,c]] = [[a,b],c] + (-1)^{\overline{|a|}\ \overline{|b|}}[b,[a,c]],$$

for any homogeneous elements $a$, $b$, and $c$ with respective degrees $|a|$, $|b|$, and $|c|$. In this definition, we use the notation $\overline{|a|} = |a| + \ell$, for an arbitrary homogeneous element $a$ of degree $|a|$.

When $[a,b] = -(-1)^{\overline{|a|}\ \overline{|b|}}[b,a]$ for all homogeneous elements $a,b \in \mathcal{A}$, then $[\cdot,\cdot]$ is a *graded Lie bracket of degree $\ell$*.

Let $\mathcal{A}$ be a graded Leibniz algebra. A *semi-representation* of $\mathcal{A}$ is a $\mathbf{k}$-module $\mathcal{M}$ endowed with a bilinear map $\varrho : \mathcal{A} \times \mathcal{M} \to \mathcal{M}$ satisfying the following property:

$$\varrho([a,b],m) = \varrho(a,\varrho(b,m)) - \varrho(b,\varrho(a,m)),$$

for all $a,b \in \mathcal{A}$ and for all $m \in \mathcal{M}$. From now on, we use the notation $\varrho(a)m$ instead of $\varrho(a,m)$.

Let $\varrho : \mathcal{A} \times \mathcal{M} \to \mathcal{M}$ be a semi-representation. Denote by $C^n(\mathcal{A}, \mathcal{M}) = \text{Hom}_k(\mathcal{A}^{\otimes n}, \mathcal{M})$ the space of k-multilinear mappings $\omega : \mathcal{A} \times \cdots \times \mathcal{A} \to \mathcal{M}$. Consider the operator $\delta : C^n(\mathcal{A}, \mathcal{M}) \to C^{n+1}(\mathcal{A}, \mathcal{M})$ defined by:

$$\delta\omega(a_1, \ldots, a_{n+1}) = \sum_{i=1}^{n+1} (-1)^{i+1} \varrho(a_i) \omega(a_1, \ldots, a_{i-1}, \widehat{a_i}, a_{i+1}, \ldots, a_{n+1})$$
$$+ \sum_{i<j} (-1)^i \omega(a_1, \ldots, \widehat{a_i}, \ldots, a_{j-1}, [a_i, a_j], a_{j+1}, \ldots, a_{n+1})$$

Then, $\delta$ satisfies $\delta \circ \delta = 0$. The corresponding cohomology spaces $HL^n(\mathcal{A}, \mathcal{M})$ are called *Leibniz cohomology groups*. See [6], [7], and [8] for more details on the theory of Leibniz cohomology.

## 3  Leibniz Algebroids

### 3.1  Definition

A *Leibniz algebroid* on a manifold $M$ consists of a vector bundle $E \to M$ with a Leibniz bracket $[\ ,\ ]$ on $\Gamma(E)$ and a bundle map $\varrho : E \to TM$, extended to a map between sections of these bundles, such that

a) $\varrho([e_1, e_2]) = [\varrho(e_1), \varrho(e_2)]$;

b) $[e_1, fe_2] = f[e_1, e_2] + (\varrho(e_1)f)e_2$,

for any $e_1$, $e_2 \in \Gamma(E)$ and for any smooth function $f$ defined on $M$. Then $\varrho$ is called the *anchor map* of the Leibniz algebroid. The vector bundle $E \to M$ is a *Lie algebroid* when the bracket $[\cdot, \cdot]$ is skew-symmetric.

### 3.2  Some examples

**Example 1.** Any Leibniz algebra of finite dimension is a Leibniz algebroid with trivial base (i.e. the base is a point).

**Example 2.** Let $\pi \in \Gamma(\Lambda^2 TM)$ be a Poisson structure on a smooth manifold $M$, that is, the Schouten-Nijenhuis bracket $[\pi, \pi]_s = 0$. Consider the vector bundle $E = T^*M \times \mathbb{R}$. Denote by $\varrho : T^*M \times \mathbb{R} \to TM$ the bundle map whose extension to sections of $E$ is given by:

$$\langle \varrho(\alpha, f),\ \beta \rangle = \pi(\alpha, \beta),$$

for any $\alpha, \beta \in \Gamma(T^*M)$, and $f \in C^\infty(M)$. Define

$$[(\alpha, f), (\beta, g)] = \left( L_{\varrho(\alpha, f)}\beta - L_{\varrho(\beta, g)}\alpha - d(\pi(\alpha, \beta)), (\varrho(\alpha, f))g \right),$$

for any $(\alpha, f), (\beta, g) \in \Gamma(E)$. The triple $(E, [\ ,\ ], \varrho)$ is a Leibniz algebroid.

**Example 3.** Let $(F, [\ ,\ ]_F, \varrho_F)$ be a Lie algebroid. Denote $E = F \times \mathbb{R}$. Consider the bracket $[\ ,\ ] : \Gamma(E) \times \Gamma(E) \to \Gamma(E)$ and the map $\varrho : \Gamma(E) \to \Gamma(TM)$ defined by:

$$[(X, f), (Y, g)] = ([X, Y]_F, \varrho_F(X)g), \quad \text{and } \varrho(X, f) = \varrho_F(X).$$

Then $(E, [\ ,\ ], \varrho)$ is a Leibniz algebroid.

**Example 4.**

A *Courant algebroid* is a vector bundle $E \to M$ equipped with a non-degenerate symmetric bilinear form $(\cdot, \cdot)$ on the bundle, a skew-symmetric bracket $[\cdot, \cdot]$ on $\Gamma(E)$ and a bundle map $\rho : E \to TM$ such that the following properties are satisfied:

(C1) $[[e_1, e_2], e_3] + c.p. = \mathcal{D}T(e_1, e_2, e_3)$;

(C2) $\rho[e_1, e_2] = [\rho e_1, \rho e_2]$;

(C3) $[e_1, f e_2] = f[e_1, e_2] + (\rho(e_1)f)e_2 - (e_1, e_2)\mathcal{D}f$;

(C4) $\rho_\circ \mathcal{D} = 0$;

(C5) $\rho(e)(h_1, h_2) = ([e, h_1] + \mathcal{D}(e, h_1), h_2) + (h_1, [e, h_2] + \mathcal{D}(e, h_2))$,

for any $e_1, e_2, e_3, e, h_1, h_2 \in \Gamma(E)$, and for any $f, g \in C^\infty(P)$. Here, the notation c.p. stands for the other terms obtained by cyclic permutations of the indices, $T$ is an $\mathbb{R}$-trilinear operation defined by:

$$T(e_1, e_2, e_3) = \frac{1}{3}([e_1, e_2], e_3) + c.p.,$$

and $\mathcal{D} : C^\infty(P) \to \Gamma(E)$ is the map defined by:

$$(\mathcal{D}f, e) = \frac{1}{2}\rho(e)f.$$

Considering the non-skew-symmetric bracket $[\cdot, \cdot]$ given by:

$$[e_1, e_2] = [e_1, e_2] + \mathcal{D}(e_1, e_2),$$

the above properties are equivalent to the following axioms:

(D1) $[e_1, [e_2, e_3]] = [[e_1, e_2], e_3] + [e_2, [e_1, e_3]]$;

(D2) $\rho[e_1, e_2] = [\rho e_1, \rho e_2]$;

(D3) $[e_1, f e_2] = f[e_1, e_2] + (\rho(e_1)f)e_2$;

(D4) $\rho(e)(h_1, h_2) = ([e, h_1], h_2) + (h_1, [e, h_2])$,

From (D1)-(D3), we see that Courant algebroids are special examples of Leibniz algebroids. See for instance [5] for more details on the theory of Courant algebroids.

**Example 5.** A *Nambu-Poisson structure* of order $p$ on an $d$-dimensional manifold $M$ (with $2 \le p \le d$) is given by a $p$-vector field $\Pi$ which satisfies the fundamental identity:

$$\{f_1, ..., f_{p-1}, \{g_1, ..., g_p\}_\Pi\}_\Pi = \sum_{k=1}^{p} \{g_1, ..., g_{k-1}, \{f_1, ..., f_{p-1}, g_k\}_\Pi, g_{k+1}, ..., g_p\}_\Pi$$

for any $f_1, \ldots, f_{p-1}, g_1, \ldots g_p \in C^\infty(M)$, where $\{\ \}_\Pi$ is defined by

$$\{f_1, \ldots, f_p\}_\Pi = \Pi(df_1, \ldots, df_p) \quad \forall f_1, \ldots, f_p \in C^\infty(M).$$

A manifold equipped with such a structure is called *Nambu-Poisson manifold*.

In [3], it is shown that every Nambu-Poisson manifold $(M, \Pi)$ has an associated Leibniz algebroid $(\Lambda^{p-1}T^*M, [\cdot, \cdot]_\Pi, \varrho_\Pi)$.

Another construction of a Leibniz algebroid arising from a Nambu-Poisson manifold is given in [2]. Also, Leibniz algebroids associated with Nambu-Jacobi structures are studied in [4].

### 3.3 Characterization of Leibniz Algebroids

Let $E$ be a vector bundle over $M$. Denote $C^0(\Gamma(E), C^\infty(M)) = C^\infty(M)$, $C^n(\Gamma(E), C^\infty(M)) = 0$, for all negative integers $n$, and

$$C^n(\Gamma(E), C^\infty(M)) = \text{Hom}_\mathbb{R}\Big( \underbrace{\Gamma(E) \times \ldots \times \Gamma(E)}_{n \text{ factors}}, \ C^\infty(M)\Big), \quad \forall n \ge 1.$$

$$C^\bullet(\Gamma(E), C^\infty(M)) = \oplus_{n \in \mathbb{Z}} C^n(\Gamma(E), C^\infty(M)).$$

By *coboundary operator* $\delta : C^\bullet(\Gamma(E), C^\infty(M)) \longrightarrow C^\bullet(\Gamma(E), C^\infty(M))$, we mean an $\mathbb{R}$-linear map of degree 1 and square zero.

**Theorem 3.1** *Let $E$ be a vector bundle over $M$. There is a one-to-one correspondence between the Leibniz algebroid structures on $E$ and the coboundary operators $\delta : C^\bullet(\Gamma(E), C^\infty(M)) \longrightarrow C^\bullet(\Gamma(E), C^\infty(M))$, which satisfy*

($P_1$) $\delta f$ *is $C^\infty(M)$-linear, for any $f \in C^\infty(M)$;*

($P_2$) $\delta$ *is a derivation on $C^\infty(M)$, i.e. $\delta(fg) = g\delta f + f\delta g$, for any $f, g \in C^\infty(M)$;*

($P_3$) $\delta\omega(e_1, fe_2) = f\delta\omega(e_1, e_2)$, *for any $\omega \in \Gamma(E^*)$, $e_1, e_2 \in \Gamma(E)$, and $f \in C^\infty(M)$.*

To prove this theorem, we need to introduce some notations and a couple of lemmas.

Given an operator $\delta$ on $C^\bullet(\Gamma(E), C^\infty(M))$, we define $\varrho$ by:

$$\varrho(e)f = (\delta f)(e),$$

for any $e \in \Gamma(E)$, $f \in C^\infty(M)$. Consider the bilinear operation $[\ ,\ ]$ given by:

$$\omega([e_1, e_2]) = \langle[e_1, e_2], \omega\rangle = \varrho(e_1)\omega(e_2) - \varrho(e_2)\omega(e_1) - \delta\omega(e_1, e_2),$$

for any $\omega \in \Gamma(E^*)$ $e_1, e_2 \in \Gamma(E)$. Then, we have:

**Lemma 3.2** *Let $\delta$ be a coboundary operator on $C^\bullet(\Gamma(E), C^\infty(M))$. Then, for any $e_1, e_2 \in \Gamma(E)$, we have*

$$\varrho([e_1, e_2]) = [\varrho(e_1),\ \varrho(e_2)].$$

*Proof:* For any $e_1, e_2 \in \Gamma(E)$, $f \in C^\infty(M)$, we have

$$\varrho([e_1, e_2])f = \langle[e_1, e_2], \delta f\rangle = \varrho(e_1)(\delta f(e_2)) - \varrho(e_2)(\delta f(e_1)) - \delta^2 f(e_1, e_2)$$
$$= \varrho(e_1)\varrho(e_2)f - \varrho(e_2)\varrho(e_1)f - \delta^2 f(e_1, e_2).$$

Since $\delta^2 = 0$, we get

$$\varrho([e_1, e_2]) = [\varrho(e_1),\ \varrho(e_2)], \quad \forall\ e_1, e_2 \in \Gamma(E).$$

∎

Now, let $i_X$ denote the interior product with respect to $X \in \Gamma(E)$, i.e. $i_X : C^n(\Gamma(E), C^\infty(M)) \longrightarrow C^{n-1}(\Gamma(E), C^\infty(M))$ is defined by:

$$i_X\omega\ (e_1, \ldots, e_{n-1}) = \omega(X, e_1, \ldots, e_{n-1}).$$

Given an operator $\delta$ on $C^\bullet(\Gamma(E), C^\infty(M))$, we define

$$L_X = i_X\delta + \delta i_X.$$

Then, we have the following lemma:

**Lemma 3.3** *Assume that $\delta$ is a coboundary operator on $C^\bullet(\Gamma(E), C^\infty(M))$. For any $X$, $Y \in \Gamma(E)$, $\omega \in C^1(\Gamma(E), C^\infty(M))$, we have*

$$L_{[X,Y]}\omega = [L_X, L_Y]\omega.$$

*Proof:* Using the fact that $i_{[X,Y]}\omega = (L_X i_Y - i_Y L_X)\omega$ and $L_X\delta = \delta L_X$, one gets

$$L_{[X,Y]}\omega = (L_X L_Y - L_Y L_X)\omega,$$

for any $X$, $Y \in \Gamma(E)$, $\omega \in C^1(\Gamma(E), C^\infty(M))$.

∎

*Proof of Theorem 3.1:* Let $(E, [\ ,\ ], \varrho)$ be a Leibniz algebroid with base $M$. We consider the degree 1 operator $\delta: C^\bullet(\Gamma(E), C^\infty(M)) \longrightarrow C^\bullet(\Gamma(E), C^\infty(M))$ given by:

$$\delta\omega(e_1, \dots, e_{n+1}) = \sum_{i=1}^{n+1}(-1)^{i+1}\varrho(e_i)\omega(e_1, \dots, e_{i-1}, \widehat{e_i}, e_{i+1}, \dots, e_{n+1})$$
$$+ \sum_{i<j}(-1)^i\omega(e_1, \dots, \widehat{e_i}, \dots, e_{j-1}, [e_i, e_j], e_{j+1}, \dots, e_{n+1})$$

for any $\omega \in C^n(\Gamma(E), C^\infty(M))$, and for any $e_1, \dots, e_{n+1} \in \Gamma(E)$. Then, $\delta$ is a coboundary operator. In general, $\delta\omega$ is not skew-symmetric (resp. $C^\infty(M)$-linear) even if $\omega$ is skew-symmetric (resp. $C^\infty(M)$-linear). But, for any $\omega \in \Gamma(E^*)$, we have

$$\delta\omega(e_1, fe_2) = \varrho(e_1)\omega(fe_2) - f\varrho(e_2)\omega(e_1) - \omega([e_1, fe_2])$$
$$= f\Big(\varrho(e_1)\omega(e_2) - \varrho(e_2)\omega(e_1)\Big) + (\varrho(e_1)f)\omega(e_2) - \omega([e_1, fe_2])$$
$$= f\delta\omega(e_1, e_2)$$

Conversely, assume that $\delta$ is a coboundary operator which satisfies Properties $(P_1)$-$(P_3)$. Define the anchor map $\varrho$ and the bracket $[\ ,\ ]$ by:

$$\varrho(e)f = (\delta f)(e), \quad \omega([e_1, e_2]) = \langle[e_1, e_2], \omega\rangle = \varrho(e_1)\omega(e_2) - \varrho(e_2)\omega(e_1) - \delta\omega(e_1, e_2)$$

for any $f \in C^\infty(M)$, $\omega \in \Gamma(E^*)$, $e_1, e_2 \in \Gamma(E)$. As a consequence of $(P_1)$-$(P_3)$, one gets:

$$\omega([e_1, fe_2]) = \varrho(e_1)\omega(fe_2) - f\varrho(e_2)\omega(e_1) - \delta\omega(e_1, fe_2)$$
$$= f\omega([e_1, e_2]) + (\varrho(e_1)f)\,\omega(e_2),$$

for any $f \in C^\infty(M)$, $\omega \in \Gamma(E^*)$, $e_1, e_2 \in \Gamma(E)$. Therefore,

$$[e_1, fe_2] = f[e_1, e_2] + (\varrho(e_1)f)e_2.$$

In view of Lemma 3.2, one has: $\varrho([e_1, e_2]) = [\varrho(e_1), \varrho(e_2)]$. Moreover, by definition

$$(L_X\omega)(Y) = ((\delta i_X + i_X\delta)\omega)(Y) = \varrho(X)\omega(Y) - \omega([X, Y]),$$

for any $X, Y \in \Gamma(E)$, and for any $\omega \in \Gamma(E^*)$. Applying this formula, we get

$$(L_{[e_1, e_2]}\omega)(e_3) = \varrho([e_1, e_2])\omega(e_3) - \omega([[e_1, e_2], e_3]),$$

and

$$\Big((L_{e_1}L_{e_2} - L_{e_2}L_{e_1})(\omega)\Big)(e_3) = \varrho([e_1, e_2])\omega(e_3) + \omega([e_2, [e_2, e_3]]) - \omega([e_1, [e_2, e_3]]),$$

for any $\omega \in \Gamma(E^*)$, $e_1, e_2, e_3 \in \Gamma(E)$.
By Lemma 3.3, we obtain

$$[e_1, [e_2, e_3]] = [[e_1, e_2], e_3] + [e_2, [e_1, e_3]],$$

for any $e_1, e_2, e_3 \in \Gamma(E)$. Thus, $E$ together with $[\ ,\ ]$, and $\varrho$ define a Leibniz algebroid. This completes the proof.

■

## 3.4 Generalized Schouten bracket

Let $(E, [\cdot, \cdot], \varrho)$ be a Leibniz algebroid over $M$. Denote by $S^2(\Gamma(E))$ the space of all $\mathbb{R}$-bilinear symmetric maps $\varphi : \Gamma(E) \times \Gamma(E) \to \Gamma(E)$. We assume that there exists a $\mathbb{R}$-linear map $\mathcal{R} : C^\infty(M) \to S^2(\Gamma(E))$ such that

$$[fX, Y] = f[X, Y] - (\varrho(Y)f)X + (\mathcal{R}(f))(X, Y).$$

This condition is satisfied by many Leibniz algebroids including those which derive from Courant algebroids and Nambu-Poisson manifolds.

**Proposition 3.4** *Denote* $\mathcal{G} = \oplus_{n\geq 1}\Gamma(\Lambda^n E)$. *Under the hypothesis above, there exists a unique $\mathbb{R}$-bilinear operation $[\![\cdot,\cdot]\!] : \mathcal{G} \times \mathcal{G} \to \mathcal{G}$, which satisfies the following properties:*

*(i)* $[\![\cdot,\cdot]\!]$ *coincides with* $[\cdot,\cdot]$ *on sections of $E$;*

*(ii)* $[\![P, Q \wedge R]\!] = [\![P, Q]\!] \wedge R + (-1)^{\bar{p}\,q} Q \wedge [\![P, R]\!];$

*(iii)* $[\![Q \wedge R, P]\!] = Q \wedge [\![R, P]\!] + (-1)^{\bar{p}\,r} + [\![Q, P]\!] \wedge R,$

*for any $P \in \Gamma(\Lambda^p E)$, $Q \in \Gamma(\Lambda^q E)$, and $R \in \Gamma(\Lambda^r E)$.*

*Proof:* Let $P \in \Gamma(\Lambda^p E)$, $Q \in \Gamma(\Lambda^q E)$, $R \in \Gamma(\Lambda^r E)$, and $S \in \Gamma(\Lambda^s E)$, where $p, q, r$, and $s$ are non-negative integers. The same result is obtained when one computes $[\![P \wedge Q, \, R \wedge S]\!]$ in two different ways. Indeed, on the one hand by Property (ii), we get

$$[\![P \wedge Q, \, R \wedge S]\!] = [\![P \wedge Q, \, R]\!] \wedge S + (-1)^{\overline{(p+q)}r} R \wedge [\![P \wedge Q, \, S]\!].$$

Applying (iii), we obtain

$$[\![P \wedge Q, \, R \wedge S]\!] = P \wedge [\![Q, \, R]\!] \wedge S + (-1)^{\bar{r}q}[\![P, \, R]\!] \wedge Q \wedge S$$
$$+ (-1)^{\overline{(p+q)}r} R \wedge P \wedge [\![Q, \, S]\!] + (-1)^{\overline{(p+q)}r + \bar{s}q} R \wedge [\![P, \, Q]\!] \wedge Q.$$

On the other hand, using first Property (iii) and then (ii), one gets:

$$[\![P \wedge Q, \, R \wedge S]\!] = P \wedge [\![Q, \, R \wedge S]\!] + (-1)^{\overline{(r+s)}q}[\![P, \, R \wedge S]\!] \wedge Q$$

$$= P \wedge [\![Q, \, R]\!] \wedge S + (-1)^{\bar{q}r} P \wedge R \wedge [\![Q, \, S]\!]$$

$$+ (-1)^{\overline{(r+s)}q}[\![P, \, R]\!] \wedge S \wedge Q + (-1)^{\overline{(p+q)}r + \bar{s}q} R \wedge [\![P, \, Q]\!] \wedge Q.$$

It follows immediately from the graded-symmetry property of the product $\wedge$ that the two results obtained are the same.

Moreover, Properties (i)-(iii) show that the bracket is local: the value of $[\![P, \, Q]\!]$ on an open subset of $M$ depends only on the values of $P$ and $Q$ on that open subset. Therefore, working with a basis $(e_1, \ldots, e_d)$ of local sections of $E$, we can express locally $[\![P, \, Q]\!]$ in terms of the $e_i$ and $[\![e_i, \, e_j]\!]$. Thus, the bracket $[\![\ , \ ]\!]$ exists and is unique. ∎

**Proposition 3.5** *Let $X \in \Gamma(E)$, $P \in \Gamma(\Lambda^p E)$, and $Q \in \Gamma(\Lambda^q E)$. Then,*

$$[\![X, \, [\![P, \, Q]\!]]\!] = [\![[\![X, \, P]\!], \, Q]\!] + [\![P, \, [\![X, \, Q]\!]]\!].$$

To prove this proposition, one can proceed by induction on the degrees of $P$ and $Q$ using Properties (ii) and (iii). The proof is straightforward and left to the reader.

**Remark.** In general, if $X_1, X_2 \in \Gamma(E)$, then $[\![X_1 \wedge X_2, \cdot]\!]$ is not a derivation with respect to the bracket $[\![\cdot, \cdot]\!]$. Indeed, a simple computation shows that

$$
\begin{aligned}
[\![X_1 \wedge X_2, &\ [\![P, Q]\!]]\!] - [\![[\![X_1 \wedge X_2, P]\!], Q]\!] - (-1)^{\overline{P}}[\![P, [\![X_1 \wedge X_2, Q]\!]]\!] \\
&= (-1)^p \Big( [\![X_1, P]\!] \wedge [\![X_2, Q]\!] + [\![P, X_1]\!] \wedge [\![X_2, Q]\!] \Big) \\
&\quad - (-1)^{\overline{q}p} \Big( [\![X_1, Q]\!] \wedge [\![X_2, P]\!] + [\![X_2, Q]\!] \wedge [\![P, X_1]\!] \Big).
\end{aligned}
$$

Since the bracket $[\![\cdot, \cdot]\!]$ is not skew-symmetric, one may have

$$
[\![X_1, P]\!] \wedge [\![X_2, Q]\!] \neq -[\![P, X_1]\!] \wedge [\![X_2, Q]\!].
$$

Hence, $[\![X_1 \wedge X_2, \cdot]\!]$ is not, in general, a derivation with respect to $[\![\cdot, \cdot]\!]$. When the original bracket defined on sections of $E$ is skew-symmetric then $[\![\cdot, \cdot]\!]$ is skew-symmetric. In this case $(\mathcal{G}, [\![\cdot, \cdot]\!], \wedge)$ is a Gerstenhaber algebra.

### 3.5   Cartan differential calculus

Let $(E, [\cdot, \cdot], \varrho)$ be a Leibniz algebroid with base $M$. For any $X \in \Gamma(E)$, we consider the standard interior product $i_X : C^n(\Gamma(E), C^\infty(M)) \longrightarrow C^n(\Gamma(E), C^\infty(M))$ and the operator $L_X : C^n(\Gamma(E), C^\infty(M)) \longrightarrow C^n(\Gamma(E), C^\infty(M))$ defined by:

$$
L_X \omega(e_1, \ldots, e_n) = \varrho(X)\omega(e_1, \ldots, e_n) - \sum_{1 \le i \le n} \omega(e_1, \ldots, e_{i-1}, [X, e_i], e_{i+1}, \ldots, e_n),
$$

One gets by simple computations the following properties:

1.  $L_X = i_X \delta + \delta i_X$;

2.  $i_{[X,Y]} = L_X i_Y - i_Y L_X$;

3.  $L_{[X,Y]} = L_X L_Y - L_Y L_X$,

for any $X, Y \in \Gamma(E)$

*3.6 Application: the modular class of a Leibniz algebroid*

In this section, we extend to the class of Leibniz algebroids the construction of the modular class of a Lie algebroid due to Evans, Lu and Weinstein (see [1]).

Consider a Leibniz algebroid $(E, [\cdot, \cdot], \varrho)$ with base $M$. An *E-connection* consists of a vector bundle $F \to M$ and an $\mathbb{R}$-bilinear map $\nabla : \Gamma(E) \times \Gamma(F) \to \Gamma(F)$ satisfying

(A1) $\nabla_X(fs) = f\nabla_X s + (\varrho(X)f) s;$

(A2) $\nabla_{fX} s = f\nabla_X s.$

In this section, we will rather use the notion of a pre-connection of $E$. Simply, a *pre-connection* of a Leibniz algebroid $E$ with base $M$ is given by a vector bundle $F \to M$ and an $\mathbb{R}$-bilinear map $\nabla : \Gamma(E) \times \Gamma(F) \to \Gamma(F)$ satisfying Axiom (A1). The curvature of $\nabla$ is defined by:

$$R(X,Y) = \nabla_X\nabla_Y - \nabla_Y\nabla_X - \nabla_{[X,Y]}, \quad \text{for any } X, Y \in \Gamma(E).$$

Using methods similar to those which are employed in [ELW], we can define the modular class of a Leibniz algebroid. Precisely, assume that $(E, [\cdot, \cdot], \varrho)$ is a Leibniz algebroid, $L$ is a real line bundle over $M$, and $\nabla : \Gamma(E) \times \Gamma(L) \to \Gamma(L)$ is a flat pre-connection (that is, the curvature of $\nabla$ is null). If $L$ has a nowhere vanishing section $s$, we define $\theta_s(X) \in C^\infty(M)$ by:

$$\nabla_X s = \theta_s(X)s.$$

By definition $\theta_s \in \text{Hom}_{\mathbb{R}}(\Gamma(E), C^\infty(M))$. A priori, $\theta_s$ is not $C^\infty(M)$-linear. Since $\nabla_{[X,Y]}s = \nabla_X\nabla_Y s - \nabla_Y\nabla_X s$, one has:

$$\theta_s([X,Y]) = \varrho(X)\theta_s(Y) - \varrho(Y)\theta_s(X),$$

for any two sections $X, Y \in \Gamma(E)$. This shows that $\theta_s$ is a 1-cocycle with respect to the coboundary operator $\delta$ associated with the Leibniz algebroid $E$. From (A1), we deduce that if $s' = fs$, for some nowhere vanishing function $f$, then

$$\theta_{s'} = \theta_s + \delta(\log|f|).$$

Hence, $[\theta_{s_1}] = [\theta_s] \in HL^1(E)$. This class is denoted by $\theta_L$ and called *characteristic class* of $E$ relative to the flat pre-connection $\nabla$.

For a non-trivializable line bundle $L$ (that is, there is not a nowhere vanishing section of $E$), the tensor product $L^2 = L \otimes L$ is a trivial line bundle.

The characteristic class of $E$ relative to a flat pre-connection $\nabla$ is:

$$\theta_L = \frac{1}{2}\theta_{L^2},$$

where $\nabla : \Gamma(E) \times \Gamma(L) \to \Gamma(L)$ is extended to sections of $L^2$ by the formula

$$\nabla_X(s_1 \otimes s_2) = (\nabla_X s_1) \otimes s_2 + s_1 \otimes \nabla_X(s_2).$$

We are going to show that there a pre-connection canonically associated with every Leibniz algebroid.

Let $(E, [\cdot, \cdot], \varrho)$ be a Leibniz algebroid with base $M$. Consider the line bundle

$$Q_E = \Lambda^{\text{top}} E \otimes \Lambda^{\text{top}} T^* M,$$

where the notation $\Lambda^{\text{top}}$ stands for the highest non-zero exterior power. We denote by $L_X$ the operator defined by:

$$L_X (Y_1 \wedge \ldots \wedge Y_n) = \sum_{i=1}^n Y_1 \wedge \ldots \wedge Y_{i-1} \wedge [X, Y_i] \wedge Y_{i+1} \wedge \ldots \wedge Y_n.$$

**Theorem 3.6** *Let $(E, \varrho, [\cdot, \cdot])$ be a Leibniz algebroid with base $M$. Then the operator $\nabla : \Gamma(E) \times \Gamma(Q_E) \to \Gamma(Q_E)$ given by:*

$$\nabla_X (u \otimes \nu) = L_X(u) \otimes \nu + u \otimes L_{\varrho(X)}\nu$$

*defines a flat pre-connection of $E$. Moreover, if $[\cdot, \cdot]$ is skew-symmetric, then $\nabla$ is a flat connection.*

*Proof:* For any $f \in C^\infty(M)$, $u \otimes \nu \in \Gamma(Q_E)$, we have

$$\begin{aligned}
\nabla_X(fu \otimes \nu) &= (L_X fu) \otimes \nu + fu \otimes L_{\varrho(X)}\nu \\
&= f\nabla_X (u \otimes \nu) + (\varrho(X)f)(u \otimes \nu)
\end{aligned}$$

Furthermore,

$$\begin{aligned}
\nabla_X \nabla_Y (u \otimes \nu) &= \nabla_X\left((L_Y u) \otimes \nu + u \otimes L_{\varrho(Y)}\nu\right) \\
&= (L_X L_Y u) \otimes \nu + L_Y u \otimes L_{\varrho(X)}\nu \\
&\quad + L_X u \otimes L_{\varrho(Y)}\nu + u \otimes L_{\varrho(X)} L_{\varrho(Y)}\nu.
\end{aligned}$$

Since $L_X L_Y - L_Y L_X = L_{[X,Y]}$ and $\varrho[X,Y] = [\varrho(X), \varrho(Y)]$, combining the terms we obtain

$$(\nabla_X \nabla_Y - \nabla_Y \nabla_X)(u \otimes \nu) = (L_{[X,Y]}u) \otimes \nu + u \otimes L_{\varrho([X,Y])}\nu = \nabla_{[X,Y]}(u \otimes \nu)$$

Therefore, $\nabla$ defines a flat pre-connection of $E$.

Now assume that $[\cdot,\cdot]$ is skew-symmetric, (i.e. $E$ is a Lie algebroid). Then,

$$[fX,Y] = f[X,Y] - (\varrho(Y)f)X.$$

Applying this property and the fact that

$$L_{f\varrho(X)}\nu = fL_{\varrho(X)}\nu + (\varrho(X)f)\nu, \quad \forall X \in \Gamma(E), \quad \forall \nu \in \Gamma(\Lambda^{\text{top}}T^*M),$$

one gets by a short computation $\nabla_{fX}u\otimes\nu = f\nabla_X u\otimes\nu$, for any $f \in C^\infty(M)$ and for any $u \otimes \nu \in \Gamma(Q_E)$. Thus, $\nabla$ becomes a flat connection when $[\cdot,\cdot]$ is skew-symmetric. ∎

This theorem generalizes Evans, Lu, and Weinstein's result (see Theorem 3.3 in [1]) and allows to introduce the following definition:

**Definition 3.7** *The modular class of a Leibniz algebroid $E$ is the cohomology class in $M_E \in HL^1(E)$, which corresponds to the flat pre-connection given by Theorem 3.6, that is, $M_E = \frac{1}{2}\theta_{Q_E}$.*

### 3.7 Relationship with the modular class of a Lie algebroid

Example 3 shows every Lie algebroid $E$ can be lifted to a Leibniz algebroid structure on $E \times \mathbb{R}$. In fact, the modular class of the original Lie algebroid [1] corresponds to the the modular class of the new Leibniz algebroid. Indeed, consider $p_n : C^n(\Gamma(E \times \mathbb{R}), C^\infty(M)) \to C^n(\Gamma(E), C^\infty(M))$ defined by

$$(p_n(\omega))(e_1,\ldots,e_n) = \omega((e_1,0), \ldots,(e_n,0)),$$

for any $\omega \in C^n(\Gamma(E \times \mathbb{R}), C^\infty(M))$, and for any $e_1,\ldots,e_n \in \Gamma(E)$. Observe that the mappings $p_n$ induce a homomorphism of complexes

$$p : (C^\bullet(\Gamma(E \times \mathbb{R}), C^\infty(M)), \delta) \to (C^\bullet(\Gamma(E), C^\infty(M)), \delta').$$

Hence, this gives a homomorphism in cohomology. In particular, if $M_{E\times\mathbb{R}}$ is the modular class of a Leibniz algebroid structure on $E \times \mathbb{R}$ obtained by lifting a Lie algebroid structure on $E$ then $p(M_{E\times\mathbb{R}})$ is the the modular class of that Lie algebroid structure. Therefore, if the modular class $p(M_{E\times\mathbb{R}})$ is non-null, so is $M_{E\times\mathbb{R}}$.

### Acknowledgements

I would like to thank A. Banyaga and J. C. Marrero for helpful comments.

## References

1. Evens S., Lu J.-H., Weinstein A., *Transverse measures, the modular class and a cohomology pairing for Lie algebroids*, Quart. J. Math. Oxford Ser. (2) 50 (1999), 417-436.
2. Hagiwara Y., *Nambu-Dirac manifolds*, J. Phys. A **35** (2002), 1263-1281.
3. Ibáñez R., de León M., Marrero J. C., Padrón E., *Leibniz algebroid associated with a Nambu-Poisson structure*, J. Phys. A **32** (1999), no. 46, 8129-8144.
4. Ibáñez R., Lopez B., Marrero J. C., Padrón E., *Matched pairs of Leibniz algebroids, Nambu-Jacobi structures and modular class* C. R. Acad. Sci. Paris Sér. I Math. 333 (2001), 861-866.
5. Liu Z.-J., Weinstein A., Xu P., *Manin triples for Lie bialgebroids*, J. Diff. Geom. **45** (1997), 547-574.
6. Loday J.-L., *Une version non commutative des algèbres de Lie: les algèbres de Leibniz*, Enseign. Math. (2) **39** (1993), 269-293.
7. Loday J.-L., Pirashvili T., *Universal enveloping algebras of Leibniz algebras and (co)homology*, Math. Ann. **296** (1993), 139-158.
8. Lodder J., *Leibniz cohomology for differentiable manifolds*, Ann. Inst. Fourier (Grenoble) **48** (1998), 73-95.
9. Kosmann-Schwarzbach Y., Magri, F., *Poisson-Nijenhuis structures*, Ann. Inst. H. Poincaré Phys. Théor. **53** (1990), 35–81.
10. Mackenzie K., Xu P., *Lie bialgebroids and Poisson groupoids*, Duke Math. J. **73** (1994), 415-452

# ON THE GEOMETRY OF LOCALLY CONFORMAL
# SYMPLECTIC MANIFOLDS

A. BANYAGA

*Department of Mathematics, The Pennsylvania State University, University Park,*
*PA 16802, USA*
*E-mail: banyaga@math.psu.edu*

Locally conformal symplectic manifolds fit into two categories: the exact and non exact ones. From the viewpoint of the Erlanger Program of Klein, the study of the non exact ones should be similar to the study of symplectic structures. The exact ones are very special: we prove a structure theorem for some of them. We show that if a compact exact lcs manifold satisfies some additionnal hypothesis, then it is fibered over $S^1$, each fiber carries a contact form, and the manifold carries a Poisson structure. Moreover, we show that on smooth manifolds carrying a locally conformal pre-symplectic form of rank at least four everywhere, the extended Lee homomorphism can be defined and globalized like in the lcs case.

## 1   Introduction and statements of the results

A symplectic form on a smooth manifold $M$ is a non-degenerate closed 2-form $\Omega$ on $M$. The existence of a symplectic form on a manifold puts strong restrictions on it. Actually the problem of existence of symplectic structures on compact manifolds is still unsolved. A manifold has a non-degenerate two form if and only if it has an almost complex structure. This is a homotopy type condition. The geometrical constraint is that the non degenerate two form be closed. It is then reasonnable to try to "relax " this condition: one comes up with the notion of *locally conformal symplectic structures*.

A *locally conformal symplectic* (lcs) form on a smooth manifold $M$ is a non-degenerate 2-form $\Omega$ such that there exists an open cover $\mathcal{U} = (U_i)$ and smooth positive functions $\lambda_i$ on $U_i$ such that

$$\Omega_i = \lambda_i(\Omega_{|U_i})$$

is a symplectic form on $U_i$ . If for all $i$ , $\lambda_i = 1$, the form $\Omega$ is a symplectic form. Lee [8] observed that the 1-forms $\{d(ln\lambda_i)\}$ fit together into a closed 1-form $\omega$ such that

$$d\Omega = -\omega \wedge \Omega. \tag{1}$$

The 1-form $\omega$ is uniquely determined by $\Omega$ and is called the Lee form of $\Omega$. The pair $(\Omega, \omega)$ of a lcs form and its Lee form will be called a lcs pair.

The uniqueness of the Lee form is a consequence of the following elemetary fact which will be often used in this paper:

**Lemma 1** *If a 2-form $\Omega$ has rank at least 4 at every point, and $\alpha$ is any 1-form, then $\alpha \wedge \Omega = 0$ implies that $\alpha$ is identically zero.*

Conversely, if a non-degenerate 2-form $\Omega$ satisfies (1), with some closed 1-form $\omega$ ( uniquely determined by $\Omega$), and $\mathcal{U} = (U)_i$ is an open cover with contractible open sets, then by Poincaré lemma, $\omega_{|U_i} = d(ln\lambda_i)$, for some positive function $\lambda_i$ on $U_i$. An immediate calculation shows that $d(\lambda_i\Omega_{|U_i}) = 0$. Since $\lambda_i\Omega_{|U_i}$ is non-degenerate, it is a symplectic form.

The same argument shows that a simply connected manifold admits a lcs form iff it admits a symplectic form.

**Definition 1** *A locally conformal pre-symplectic form (lcps) on a smooth manifold $M$ is a 2-form $\Omega$ such that there exists a closed 1-form $\omega$ satisfying (1).*

**Remarks**
If $\Omega$ is a lcs form on a smooth manifold $M$, the restriction to any submanifold $N$ of $M$ of $\Omega$ is a lcps form. Hence this definition may be too general and one may wish to restrict the nature of the allowed submanifolds. The most interesting definition seems to be the following:

**Definition 2** *A locally conformal pre-symplectic form (lcps) on a smooth manifold $M$ is a 2-form $\Omega$ of constant rank such that there exists a closed 1-form $\omega$ satisfying (1).*

In this paper, we consider an intermediate situation, in which we ask, that **the lcps forms $\Omega$ ( Definition 1), have a rank which is at least four everywhere: i.e. that $(\Omega(x))^2 \neq 0$ for all $x \in M$.**

In that case, Lemma 1 implies that $\omega$ (in Definition 1) is uniquely determined by the equation (1).We will then continue to call $\omega$ the Lee form of $\Omega$ and refer to $(\Omega, \omega)$ as a lcps pair.

Two lcps forms $\Omega$, $\Omega'$ on a smooth manifold $M$ are said to be (conformally) equivalent if $\Omega' = f\Omega$ , for some positive function $f$ on $M$. If $\omega$ is the Lee form of $\Omega$, then $\omega' = \omega - dlnf$ is the Lee form of $\Omega'$. The lcps pairs $(\Omega, \omega)$ and $(\Omega', \omega')$ are said to be equivalent.

A locally conformal pre-symplectic (lcps) structure $\mathcal{S}$ on a smooth man-

ifold $M$ is an equivalence class of lcps forms. The couple $(M, S)$ is called a
*lcps manifold.*

If $\Omega$ is a representative of $S$, we write $\Omega \in S$, or $S = [\Omega]$. If $\omega = 0$ in
the definition above, then $\Omega$ is a pre-symplectic form. In that case the lcps
structure $S$ is said to be a *global conformal pre-symplectic* (gcps) structure
and we write $S = \mathcal{O}$. A lcps structure $S$ is global iff for any $\Omega \in S$, the
corresponding Lee form $\omega$ is exact.

The connection with (pre) symplectic geometry is given by the following.

**Theorem 1 (4)** *Let $\Omega$ be a lcs ( lcps) form on a smooth manifold $M$ with
Lee form $\omega$. Let $p : \tilde{M} \to M$ be the minimum cover associated with $\omega$ (whose
group of automorphisms is the group of periods of $\omega$). Let $\lambda$ be a positive
function on $\tilde{M}$ such that $p^*\omega = d(ln\lambda)$, then $\tilde{\Omega} = \lambda p^*\Omega$ is a symplectic form
( a pre-symplectic form) on $\tilde{M}$, and its class $[\tilde{\Omega}]$ depends only on the lcs
structure $[\Omega]$.*

Observe that we may in Theorem 1 consider instead the universal covering
of $M$. However since the universal cover is simply connected, the Calabi
invariants of $\tilde{\Omega}$ will be zero. See [1,[3],[4]]. Hence the universal cover is not
appropriate to study the (pre) symplectic geometry of $(\tilde{M}, \tilde{\Omega})$.

For any closed 1-form $\omega$ on a smooth manifold $M$, the operator $d_\omega$ which
assigns to a p-form $\gamma$ the (p+1)-form

$$d_\omega \gamma = d\gamma + \omega \wedge \gamma \tag{2}$$

is a coboundary operator, i.e. $d_\omega \circ d_\omega = 0$.

The cohomology of differential forms with this coboundary operator will
be denoted by $H_\omega^*(M)$ and will be called the $d_\omega$ - cohomology. It was introduced by Vaisman [11] and by Guerida-Lichnerowicz [6] as a tool to the study
of lcs structures. This cohomology will be developped in full details in [5].

A lcps form $\Omega$ is a $d_\omega$ closed 2-form ( where $\omega$ is the Lee form). Hence a
lcps form $\Omega$ with Lee form $\omega$ defines an element $[\Omega] \in H_\omega^2(M)$. If this class is
zero, we say that $\Omega$ is $d_\omega$-exact.

**Proposition 1** *If two lcps pairs $(\Omega, \omega)$ and $(\Omega', \omega')$ are equivalent, then $\Omega$ is
$d_\omega$-exact iff $\Omega'$ is $d_{\omega'}$-exact.*

Hence we define a lcps structure $S$ to be an **exact structure** if it admits
a $d_\omega$ exact representative. Otherwise the structure will be called **non exact**.

In [7], it is shown that the infinite dimensional Lie group $Diff_S(M)$ of automorphisms of a lcs structure $S$ or its infinitesimal version $\mathcal{X}_S(M)$ determines the lcs structure $S$. This generalizes the corresponding fact in symplectic geometry [1].

The group of automorphisms $Diff_S(M)$ of a lcps structure $S$ on a smooth manifold $M$ is the set of all diffeomorphisms $\phi$ of $M$ such that $\phi^*\Omega = f_\phi\Omega$, where $f_\phi$ is a smooth function on $M$, and $\Omega$ is any representative of $S$. We will denote by $Diff_S(M)_0$ the subgroup formed by those $\phi \in Diff_S(M)$, which are isotopic to the identity through $Diff_S(M)$.

The Lie algebra $\mathcal{X}_S(M)$ of infinitesimal automorphisms of $S$, consists of vector fields $X$ on $M$ such that $L_X\Omega = \rho_X\Omega$, where $\rho_X$ is a smooth function on $M$. Here $L_X$ stands for the Lie derivative in the direction $X$. We denote by $\mathcal{X}_\Omega(M)$ the Lie subalgebra of vector fields $X$ such that $L_X\Omega = 0$.

The following result is another advocate of Klein's Erlanger Program creed that **"properties of a geometry should be determined by its automorphism group."**

**Theorem 2** *Let $(M, S)$ be a connected lcps manifold, and let $\Omega$ be a lcps form (of rank at least 4 everywhere) representing $S$, with Lee form $\omega$.*

*1.If a vector field $X$ on $M$ is an infinitesimal automorphism of $S$, there is a constant $l_X$ such that*

$$d_\omega\theta = l_X\Omega \tag{3}$$

*where $\theta = i(X)\Omega$.*
*Hence the form $\Omega$ is $d_\omega$ exact if $l_X \neq 0$.*
*If $L_X\Omega = \rho_X\Omega$, then*

$$l_X = \rho_X + \omega(X). \tag{4}$$

*2. The assignement $X \mapsto l_X$ is a Lie algebra homomorphism $l : \mathcal{X}_S(M) \to \mathbf{R}$ depending only on the conformal class $S$ of $\Omega$, (called the extended Lee homomorphism).*

*Let us now assume that $\Omega$ is a lcs form and that $M$ is compact. We have:*
*3. If $\Omega$ is normalized so that the volume $V = \int_M (\Omega)^n = 1$, then*

$$l_X = \int_M \omega(X)(\Omega)^n \tag{5}$$

*4. If $l_X \neq 0$ and $M$ is compact, then the structure $S$ is not global. Hence on a compact manifold, a global lcs structure is never exact.*

   5. *Suppose that $M$ is compact, $l_X \neq 0$, and $d\theta$ has constant rank. Then $M$ is fibered over $S^1$ and the restriction of $\theta$ to each fiber is a contact form. Moreover $M$ carries a Poisson structure.*

The following fact is a consequence of the proof of Theorem 2

**Proposition 2** *Let $(M, \mathcal{S})$ be a compact and connected lcs manifold, and let $\Omega$ be a lcs form representing $\mathcal{S}$. If an automorphism $X$ of $\mathcal{S}$ satisfies $l_X \neq 0$, and $d\theta$ has constant rank, where $\theta = i(X)\Omega$, then $X$ preserves $\Omega$, i.e. $L_X\Omega = 0$.*

It follows that lcs forms satisfying the hypothesis of part (5) in Theorem 2 are precisely Vaisman's [10] lcs **of the first kind**.

Statements 1 and 2 are well known facts in lcs geometry and can be found in [6], [10], see also [4]. The proofs in the more general case of lcps structures are analogous. The point is that Lemma 1 works not only for non-degenerate two forms, but more generally for 2-forms of rank at least four everywhere. For completeness, we give the proofs.

**Proposition 3** *A lcs structure $\mathcal{S}$ is exact iff its extended Lee homomorphism is non trivial.*

Note that from the fact that a lcps is exact, we can not conclude that $l$ is non trivial. Indeed the mapping assigning to $X$ the 1-form $i(X)\Omega$ is not surjective.

The extended Lee homomorphism has the following geometrical interpretation:

**Theorem 3** *Let $(M, \mathcal{S})$ be a lcps manifold and $X \in \mathcal{X}_\mathcal{S}(M)$. Let $\tilde{X}$ be the lift of $X$ to the cover $\tilde{M}$ in Theorem 1, then $L_{\tilde{X}}\tilde{\Omega} = l(X)\tilde{\Omega}$.*

**Corollary 1** *All the infinitesimal automorphisms of a lcs structure $[\Omega]$ lift to symplectic automorphisms of $(\tilde{M}, \tilde{\Omega})$ iff the lcs $[\Omega]$ is not exact.*

Hence, from the viewpoint of the Erlanger Program, this corollary implies that the study of non exact lcs structures should follow from that of symplectic manifolds. Indeed, the properties of such lcs manifolds are encoded in the Lie algebra of its infinitesimal automorphisms [7], which appear as a sub-algebra of the automorphisms of the corresponding symplectic structure, which encodes its geometry [1].

## 1.1 Integration of the extended Lee homomorphism

In the case of a compact lcs manifold $(M, \Omega)$, it is possible to "analytically" integrate the homomorphism $l : \mathcal{X}_S(M) \to \mathbf{R}$ to a homomorphism $\mathcal{L} : Diff_S(M)_0 \to \mathbf{R}/\Delta$, where $\Delta$ is a subgroup of $\mathbf{R}$ [4]. If $M$ is connected and non compact, the analytic methods do not work (one simple reason is that vector fields with unrestricted supports are not complete) and the restriction of $l$ to automorphisms with compact support is identically zero.

A geometric integration of the homomorphism $l$, which works for compact manifolds as well as for connected non compact manifolds, was given in [3], [4]. The same arguments extend to the lcps case:

**Theorem 4** *Let $(M, \mathcal{S})$ ba a connected lcps manifold, a lcps form $\Omega \in \mathcal{S}$, with Lee form $\omega$, where the rank of $\Omega$ is at least four at each point. Let $\pi : \tilde{M} \to M$ the covering associated with $\omega$, a function $\lambda : \tilde{M} \to \mathbf{R}$ such that $\pi^*\omega = d(ln\lambda)$.*

*For each $\phi \in Diff_S(M)_0$, let $\tilde{\phi} : \tilde{M} \to \tilde{M}$ be a diffeomorphism covering $\phi$, i.e such that $\pi \circ \tilde{\phi} = \phi \circ \pi$, then*

$$\frac{\lambda \circ \tilde{\phi}}{\lambda}.(f_\phi \circ \pi) \tag{6}$$

*is a non zero constant $b_{\tilde{\phi}}$, independent of the choice of $\lambda$.*

*If $\hat{\phi}$ is another lifting of $\phi$, then $b_{\hat{\phi}} = \sigma.b_{\tilde{\phi}}$, where $\sigma$ belongs to the group $\Delta$ ( subgroup of $\mathbf{R}$) of periods of $\omega$.*

*The correspondance : $\phi \mapsto b_{\tilde{\phi}}$ is a well defined group homomorphism:*

$$\mathcal{L} : Diff_S(M)_0 \to \mathbf{R}^+/\Delta \tag{7}$$

*which does not depend on the choice of $(\Omega, \omega) \in \mathcal{S}$, i.e. is a conformal invariant.*

*The number $b_{\tilde{\phi}}$ is the similitude ratio of $\tilde{\phi}$ , i.e. $\tilde{\phi}^*\tilde{\Omega} = b_{\tilde{\phi}}\tilde{\Omega}$. Hence the Kernel $G$ of $\mathcal{L}$ is a normal subgroup which can be identified with a quotient of a connected subgroup of the group of (pre)-symplectic diffeomorphisms of $(\tilde{M}, \tilde{\Omega})$.*

*Let $\phi_t$ be the local 1-parameter group of diffeomorphisms generated by an infinitesimal automorphism $X \in \mathcal{X}_S(M)$, then:*

$$\frac{d}{dt}(ln(b_{\tilde{\phi}_t}))|_{t=0} = l(X) \tag{8}$$

The proof of Theorem 4 is the same as in [4] once we have taken note of the following:

**Proposition 4** *Let $\Theta$ be a closed 2-form of rank at least four everywhere on a smooth connected manifold $M$. If $\phi : M \to M$ is a smooth diffeomorphism such that $\phi^*\Theta = f\Theta$ for some function $f$, then $f$ is a constant.*

This is an immediate consequence of Lemma 1. Indeed : $0 = \phi^*(d\Theta) = d(\phi^*\Theta) = df \wedge \Theta$. By Lemma 1, $df = 0$ □

## 2 Examples of exact lcs and of lcps manifolds

Example 1: Let $\alpha$ be a contact form on a $(2n+1)$-dimensional manifold $N$: that is a 1-form $\alpha$ such that $\alpha \wedge (d\alpha)^n$ is a volume form. Let $M = N \times S^1$ be the cartesian product of $N$ and the circle $S^1$. Let $p : M \to N$ and $q : M \to S^1$ the projections on each factor. Let $\mathcal{L}$ be the length form on $S^1$ : $\mathcal{L} = xdy - ydx$. Then :

$$\Omega = d\theta + \omega \wedge \theta$$

where $\theta = p^*\alpha$ and $\omega = q^*\mathcal{L}$, is a lcs form.

Morover if $\xi$ is the vector field on $S^1$, $\quad (x,y) \mapsto (-y,x)$, the vector field $X = (0,\xi)$ on $M$, satisfies : $\omega(X) = 1, L_X\theta = 0$ and hence $L_X\Omega = 0$. Therefore $l_X = 1$.

This example can be generalized as follows:

Example 2: Suspension of strictly contact diffeomorphisms.
Let $(N, \alpha)$ be a contact manifold, and $h : N \to N$ a smooth diffeomorphism such that $h^*\alpha = \alpha$. Let $M = N_h$ be the suspension ( mapping torus) of $h$, i.e.

$$N_h = N \times [0,1]/\{(x,0) \approx (h(x),1)\}.$$

This is a fibration $\pi : M \to S^1$ with fibers diffeomorphic with $N$. Let $p : N \times [0,1] \to N$ be the projection $(x,t) \mapsto x$, then the 1-form $p^*\alpha$ descends to the quotient : so we get a 1-form $\theta$ on $M$, the restriction of which to each fiber is a contact form. If $\omega = \pi^*\mathcal{L}$, then $\Omega = d\theta + \omega \wedge \theta$ is a lcs form , and if $X$ is the vector field defined by $i(X)\Omega = \theta$, then $l_X = 1$.

Example 3. Flat $S^1$ principal bundles over contact manifolds[10].
Let $\pi : M \to B$ be a flat principal $S^1$ bundle over a contact manifold $B$, with contact form $\alpha$. Let $\omega$ be a connection form on $M$ : this is an ordinary 1-form on $M$, which is closed since the bundle is flat. Then:

$$\Omega = d\theta + \omega \wedge \theta$$

where $\theta = \pi^*\alpha$, is a lcs form. Let $X$ be the horizontal lift of the Reeb field of $\alpha$, then $L_X\Omega = 0$ and $\omega(X) = 1$. Hence $l_X = 1$.

Example 4 : Submanifolds of lcs manifolds.

Let $j : N \to M$ be an embedding of a submanifold $N$ into a lcs manifold $(M, S')$, and let $(\Omega', \omega')$ be a lcs pair representing $S'$. Then $(\Omega, \omega)$, where $\Omega = j^*\Omega'$ and $\omega = j^*\omega'$, is a lcps pair ( Dfinition 1).

There are several interesting cases:

(1) The rank of $\Omega$ is constant (Definition 2).

(2) The rank of $\Omega$ is everywhere greater or equal to four ( those are the ones considered here, for which the extended Lee homomorphism of Theorem 2 is defined).

(3) Submanifolds on which $\omega = j^*\omega'$ is exact, or zero, i.e. the relative de Rham class of $\omega$ in $H^1(M, N, \mathbf{R})$ is zero.

Observe that if $\Omega'$ is $d_{\omega'}$ exact, then $\Omega$ is $d_\omega$ exact.

To get examples of these cases, we may start with a smooth map $f : M \to \mathbf{R}^k$ from a lcs manifold $(M, S')$ to $\mathbf{R}^k$ and let $j : N \to M$ be the inclusion of a regular level $N = f^{-1}(a)$, where $a$ is a regular value of $f$.

In case (2), the equation (1) says that $\Omega = 0$ defines a foliation $\mathcal{F}$. If this foliation is simple, we get a lcs structure on $P = N/\mathcal{F}$.

## 3    Proofs of the results

### 3.1    Proof of Lemma 1

Suppose there is a point $x \in M$ so that $\alpha_x \neq 0$. Complete $\alpha_x$ into a basis of the cotangent space $T_x^*(M)$ : $(\alpha_x, \theta_2, .., \theta_m)$, $m = dimM$ and write $\Omega_x$ as :

$$\Omega_x = \sum_{j=2}^{m} u_j \alpha_x \wedge \theta_j + \sum_{i<j} v_{ij} \theta_i \wedge \theta_j$$

We have:

$$0 = \alpha_x \wedge \Omega_x = \sum_{i<j} v_{ij} \alpha_x \wedge \theta_i \wedge \theta_j.$$

Since $\alpha_x \wedge \theta_i \wedge \theta_j$ are linearly independent, $v_{ij} = 0$ for all i and j. Hence $\Omega_x = \alpha_x \wedge \beta$ with $\beta = \sum_{j=2}^{m} u_j \theta_j$. Consequently, $\Omega_x^2 = 0$ contradicting the fact that the rank of $\Omega$ is assumed to be at least 4 at every point, i.e. $\Omega_x^2 \neq 0$.
□

### 3.2 Proof of Proposition 1

Let $(\Omega, \omega)$ and $(\Omega', \omega')$ be two equivalent lcps pairs. Then $\Omega' = \lambda\Omega$ and $\omega' = \omega - d(\ln\lambda)$. Suppose that $\Omega = d\theta + \omega \wedge \theta$. Then $\Omega' = \lambda(d\theta) + \omega \wedge \lambda\theta = d(\lambda\theta) - d\lambda \wedge \theta + \omega \wedge \lambda\theta = d(\lambda\theta) + (\omega - ((d\lambda)/\lambda) \wedge (\lambda\theta) = d(\lambda\theta) + (\omega - d\ln\lambda) \wedge (\lambda\theta) = d_{\omega'}(\lambda\theta)$. □

### 3.3 Proof of Theorem 2 : Part 1

Let $\Omega$ be a lcps form on a smooth manifold $M$. Given a 1-form $\theta$ on $M$, under what conditions the vector field $X$ defined by: $i(X)\Omega = \theta$ is an infinitesimal automorphism of the lcps structure $S$ defined by $\Omega$?

Let us compute $L_X\Omega$: $L_X\Omega = di(X)\Omega + i(X)(d\Omega) = d\theta + i(X)(-\omega \wedge \Omega) = d\theta - \omega(X)\Omega + \omega \wedge \theta = (d\theta + \omega \wedge \theta) - \omega(X)\Omega = d_\omega\theta - \omega(X)\Omega$.

The equation: $L_X\Omega = d_\omega\theta - \omega(X)\Omega$ shows that $X$ is an automorphism iff

$$d_\omega\theta = l_X\Omega$$

for some function $l_X$ on $M$. In that case, we have $L_X\Omega = (-\omega(X) + l_X)\Omega$. Hence the function $l_X$ satisfies:

$$l_X = \rho_X + \omega(X).$$

if

$$L_X\Omega = \rho_X\Omega$$

We need the following:

**Proposition 5** *Let $(M, S)$ be a connected lcps manifold. For any $X \in \mathcal{X}_S(M)$, the function $l_X$ is a constant function.*

**Proof**
$$0 = (d_\omega)^2\theta = d(l_X\Omega) + \omega \wedge (l_X)\Omega = dl_X \wedge \Omega + l_X(-\omega \wedge \Omega) + \omega \wedge (l_X)\Omega = dl_X \wedge \Omega.$$

By Lemma 1, $dl_X = 0$. Hence $l_X$ is a constant function. □

**Corollary 2** *Let $(M, S)$ be a connected lcps manifold, and $X \in \mathcal{X}_S(M)$. Either $\theta = i(X)\Omega$ is $d_\omega$ closed, and hence defines an element $[X] \in H^1_\omega(M)$ or $\Omega$ is $d_\omega$-exact, i.e $[\Omega] = 0 \in H^2_\omega(M)$.*

If $(\Omega, \omega)$ is a lcs pair, then $\Omega$ is $d_\omega$ exact, and hence $S$ is an exact structure, iff there is $X \in \mathcal{X}_S(M)$ with $l_X \neq 0$.

**Proof**

If $l_X = a \neq 0$, then $\Omega = d\theta' + \omega \wedge \theta'$ with $\theta' = \theta/a$ .

If now $(\Omega, \omega)$ is a lcs pair and $\Omega = d\theta + \omega \wedge \theta$, Let $X$ be defined by $i(X)\Omega = \theta$, then:
$$L_X\Omega = d\theta + i(X)d\Omega = d\theta + i(X)(-\omega \wedge \Omega) = d\theta - \omega(X)\Omega + \omega \wedge \theta = \Omega - \omega(X)\Omega = (1 - \omega(X))\Omega.$$

Hence we see that $L_X\Omega = \delta_X\Omega$ with $\delta_X = 1 - \omega(X)$. We therefore see that $X \in \mathcal{L}_S(M)$ and $l_X = 1$. □

This proves part 1 of Theorem 3 and Proposition 3.

## 3.4 Proof of part 3

We have : $L_X(\Omega)^n = nL_X\Omega \wedge (\Omega)^{n-1} = n\rho_X(\Omega)^n = n(l_X - \omega(X))(\Omega)^n$. We thus see that $d(i(X)(\Omega)^n) = n(l_X - \omega(X))(\Omega)^n$.

If $M$ is compact, Stokes formula gives: $\int_M l_X\Omega^n = \int_M \omega(X)\Omega^n$. Since $l_X$ is a constant, it follows that if $\int_M \Omega^n$ is set to be 1, that $l_X = \int_M \omega(X)\Omega^n$. □

## 3.5 Proof of part 4

We saw that if $l_X \neq 0$, then $\Omega = d\theta + \omega \wedge \theta$.

If there exists a positive function $\lambda$ such that $\lambda\Omega$ is a symplectic form, then $\omega = d(ln\lambda)$.

Therefore: $\lambda\Omega = \lambda d\theta + \lambda\omega \wedge \theta = \lambda d\theta + \lambda d(ln\lambda) \wedge \theta = \lambda d\theta + (d\lambda) \wedge \theta = d(\lambda\theta)$.

This is an exact symplectic form on a compact manifold. No symplectic form on a compact manifold is exact. Therefore $[\Omega]$ is not global □

## 3.6 Proof of part 5

Without loss of generality, we may assume that $l_X = 1$. By part 1, $\Omega = d\theta + \omega \wedge \theta$. We have:

$$\Omega^n = (d\theta)^n + n(\omega \wedge \theta \wedge (d\theta)^{n-1})$$

Since $M$ is compact, $(d\theta)^n$ must have a zero: otherwise $d\theta$ would be an exact symplectic form on a compact manifold, an impossibility. Therefore since we assumed that $d\theta$ has constant rank, it must have rank strictly less than $2n$, i.e. $(d\theta)^n$ vanishes identically. We then have that $\omega \wedge \theta \wedge (d\theta)^{n-1})$ is everywhere non zero. In particular $\omega$ and $\theta$ are non singular.

The restriction of $\theta$ to each leaf of the foliation $\mathcal{F}_\omega$ defined by $\omega = 0$ is a contact form: indeed, if $v_2, .., v_{2n}$ are linearly independent vectors tangent to $\mathcal{F}_\omega$ at $x$, then $(\theta \wedge (d\theta))^{n-1}(v_2, .., v_{2n}) \neq 0$ since for any vector $u \in T_x M$, transverse to $\mathcal{F}_\omega$,

$$\Omega^n(x)(u, v_2, ...v_{2n}) = n\omega(u)(\theta \wedge (d\theta)^{n-1}(v_2, .., v_{2n}) \neq 0.$$

In [9], Tischler approximates the 1-form $\omega$ by $q\hat{\omega}$ for some integer $q$, and $\hat{\omega}$ is a 1-form with integral periods. Basically as a consequence of the classical fact that $H^1(M, \mathbf{Z}) \approx [\mathbf{M}, \mathbf{S^1}]$, where $[M, S^1]$ is the set of homotopy classes of maps into the circle, he deduces that there is a smooth map $f : M \to S^1$ such that $\hat{\omega} = f^*(\mathcal{L})$, where $\mathcal{L}$ is the length form on $S^1$. The form $q\hat{\omega}$, and ( hence $\hat{\omega}$ ) has no zero since it is close to $\omega$; thus $f^*(\mathcal{L})$ has no zero, meaning that $f$ is a submersion. Since $M$ is compact, it is a fibration. The fibers are defined by the equation $\hat{\omega} = 0 = q\hat{\omega}$, which is close to the equation $\omega = 0$ defining $\mathcal{F}_\omega$. The fibers of $\pi$ and the leaves of the foliation $\mathcal{F}_\omega$ are thus arbitrarily $C^\infty$ close: this means that their tangent planes ( which are defined by the equations $\omega = 0$ and $\hat{\omega} = 0$) are $C^\infty$ close. They are considered as space of sections of Grassmanian bundles, with the $C^\infty$ topology.

This implies that the restriction of $\theta$ to the fibers of $\pi$ is a contact form. Indeed, let $x$ be a point in a fiber $F$ of $\pi$, and let $L$ be the leaf of $\mathcal{F}_\omega$ through $x$. Using normal coordinates in $M$ (for some riemannian metric on $M$), we get a diffeomorphism $\phi$ from a neigborhood $D_L$ of $x$ in $L$ to a neigborhood $D_F$ of $x$ in $F$, such that $\phi(x) = x$, and $d_x\phi$ is arbitrarly close to the identity. Hence if $\mathcal{B} = (v_2, ..., v_{2n})$ is a basis of $T_x L$ such that $(\theta \wedge (d\theta)^{n-1})(v_2, .., v_{2n}) \neq 0$, then $(\theta \wedge (d\theta))^{n-1}(w_2, .., w_{2n}) \neq 0$, where $w_i = d_x\phi(v_i)$. Hence the restriction of $\theta$ is a contact form on $F$.

The following equations:

$$\omega(V) = 1, \quad i(V)\theta = i(V)d\theta = 0$$

define uniquely a vector field $V$ such that $L_V\Omega = 0$ [10]. We have : $i(V)\Omega = i(V)(d\theta + \omega \wedge \theta) = \theta = i(X)\Omega$. Since $\Omega$ is non degenerate, $X = V$. Our original vector $X$ actually satisfied : $L_X\Omega = 0$ and hence $\omega(X) = 1$, since $l_X = 1$.

This fact is a proof for Proposition 2.

Let $Z$ be the unique vector field defined by $i(Z)\Omega = \omega$. We have : $L_Z\Omega = L_Z\omega = 0$. Therefore $i([Z, X])\Omega = i(Z)L_X\Omega - L_X i(Z)\Omega = 0$. Hence $[Z, X] =$

0. Therefore, $\{X, Z\}$ span a 2-dimensional foliation $\mathcal{F}$, called the "vertical foliation" [10].

We have:

$$\Omega(X, Z) = i(X)(i(Z)\Omega)) = i(X)\omega = 1.$$

Therefore the restriction $\Omega_S$ of $\Omega$ to each leaf $S$ of $\mathcal{S}$ is a symplectic form, i.e. $\mathcal{S}$ is a symplectic foliation. The Dirac bracket $\{,\}_D$ on $C^\infty(M)$ defines a Poisson structure $\Lambda_D$ on $M$ the usual way: $\Lambda_D(df, dg) = \{f, g\}_D$ for all functions $f$ and $g$.

Recall that this is defined as follows : let $f$ be a smooth function, for each leaf $S$, we define a vector field $X_{f,S}$ by : $i(X_{f,S})(\Omega_S) = d(f_{|S})$. These vector fields fit together into a smooth vector field $X_f$. The Dirac bracket is defined as $\{f, g\}_D = X_f \cdot g$. This generalizes the example 3.5 in [12]. □

### 3.7 Proof of part 2 of Theorem 2

It is obvious that the map $l$ is a vector space homomorphism. To show it is a Lie algebra homomorphism, we only need to show that it vanishes on brackets of vector fields in $\mathcal{X}_S(M)$. For $X, Y \in \mathcal{X}_S(M)$, we have:

$$L_{[X,Y]}\Omega = \delta_{[X,Y]}\Omega$$

$$= L_X L_Y \Omega - L_Y L_X \Omega = (X \cdot \delta_Y - Y \cdot \delta_X)\Omega$$

Hence, $\delta_{[X,Y]} = (X \cdot \delta_Y - Y \cdot \delta_X)$.
On the other hand, since $\omega$ is a closed 1-form:

$$\omega([X, Y]) = L_X(i(Y)\omega - i(Y)L_X\omega = X \cdot \omega(Y) - Y \cdot \omega(X).$$

Moreover, for any pair of vector fields $U, V, \ V \in \mathcal{X}_S(M)$, we have:

$$U \cdot \delta_V = -U \cdot \omega(V)$$

since $\delta_V = -\omega(V) + l(V)$ and $l(V)$ is constant. Therefore

$$l([X, Y]) = \delta_{[X,Y]} + \omega([X, Y]) = ((X \cdot \delta_Y - Y \cdot \delta_X)) + ((X \cdot \omega(Y) - Y \cdot \omega(X))) = 0$$

To show that $l$ is an invariant of $[\Omega]$, consider $X$ an automorphism: $L_X\Omega = \rho_X\Omega$ and $L_X\Omega' = \rho'_X\Omega'$, where $\Omega' = \lambda\Omega$, for some positive function $\lambda$. We know that the Lee form $\omega'$ of $\Omega'$ is $\omega - d(ln\lambda)$. A direct calculation shows that $\rho'_X = d(ln\lambda)(X) + \rho_X$. Hence $\omega'(X) + \rho'_X = (\omega(X) - dln\lambda(X)) + (dln\lambda(X) + \rho_X) = \omega(X) + \rho_X$.

□

# References

1. A. Banyaga, *The structure of classical diffeomorphism groups*, Mathematics and its Applications Vol 400, Kluwer Academic Publisher, Dordrecht, The Netherlands, 1997,
2. A. Banyaga, *Some properties of locally conformal symplectic structures*, Comment. Math. Helvetici, to appear.
3. A. Banyaga, *Quelques invariants des structures localement conformément symplectiques*, C.R.Acad. Sci. Paris t 332 , Serie 1 (2001) 29-32.
4. A. Banyaga, *A geometric integration of the extended Lee homomorphism*, Journal of Geometry and Physics, 39(2001) 30-44.
5. A. Banyaga, *Introduction to locally conformal symplectic geometry*, In preparation.
6. F. Guerida and A. Lichnerowicz, *Géometrie des algèbres de Lie locales de Kirillov*, J.Math. Pures et Appl. 63(1984), 407-484.
7. S. Haller and T. Rybicki, *On the group of diffeomorphisms preserving a locally conformal symplectic structure*, Ann. Global Anal. and Geom. 17 (1999) 475-502.
8. H.C. Lee, *A kind of even-dimensional differential geometry and its application to exterior calculus*, Amer. J. Math. 65(1943) pp 433-438.
9. D. Tischler, *On fibering certain manifolds over $S^1$*, Topology 9(1970), 153-154.
10. I. Vaisman, *Locally conformal symplectic manifolds*, Inter. J. Math. and Math. Sci. Vol 8 no 3 (1983), p 521-536.
11. I. Vaisman, *Remarkable operators and commutation formulas on locally conformal kaehler manifolds*, Compositio Math. , 40(1980) 287-299.
12. I. Vaisman, *Lectures on the geometry of Poisson manifolds*, Progress in Math, no 118. Birkhauser, 1994.

# SOME PROPERTIES OF LOCALLY CONFORMAL SYMPLECTIC MANIFOLDS

STEFAN HALLER

*Institute of Mathematics, University of Vienna*
*Strudlhofgasse 4, A-1090 Vienna, Austria*
*E-mail: stefan@mat.univie.ac.at*

The aim of this note is to give an overview of what locally conformal symplectic manifolds are and to present some results, well known from symplectic geometry, which can be generalized to this slightly larger category.

## 1  Locally conformal symplectic manifolds

A locally conformal symplectic manifold, l.c.s. for short, is a triple $(M, \Omega, \omega)$, where $M$ is a finite dimensional smooth manifold, $\Omega$ is a non-degenerate 2-form on $M$ and $\omega$ is a closed 1-form on $M$, such that $d^\omega \Omega = 0$, where

$$d^\omega : \Omega^*(M) \to \Omega^{*+1}(M), \quad d^\omega \alpha := d\alpha + \omega \wedge \alpha$$

is the Witten deformed differential. We will always assume, that $M$ is connected and $\partial M = \emptyset$.

*Remark 1.* The dimension of an l.c.s. manifold has to be even. Since $\Omega^n$ is nowhere vanishing an l.c.s. manifold possesses a canonical orientation.

*Remark 2.* If $\dim(M) > 2$ then $\omega$ is uniquely determined by $\Omega$. Indeed if $d^\omega \Omega = d^{\omega'} \Omega = 0$ and $\omega_x \neq \omega'_x$ at a point $x \in M$, then $(\omega'_x - \omega_x) \wedge \Omega_x = 0$ and hence $\Omega_x = (\omega'_x - \omega_x) \wedge \alpha$ for some $\alpha \in T_x^* M$, but this contradicts the non-degeneracy of $\Omega$ at the point $x$.

*Remark 3.* If $(\Omega, \omega)$ is an l.c.s. structure on $M$ and $a \in C^\infty(M, \mathbb{R})$ then $(e^a \Omega, \omega - da)$ is again an l.c.s. structure on $M$. Two l.c.s. structures $(\Omega, \omega)$ and $(\Omega', \omega')$ on $M$ are called conformally equivalent if $(\Omega', \omega') = (e^a \Omega, \omega - da)$ for some $a \in C^\infty(M, \mathbb{R})$. We will write $(\Omega, \omega) \sim (\Omega', \omega')$ in this case. Note that the canonical orientation does only depend on the conformal equivalence class of $(M, \Omega, \omega)$.

*Remark 4.* Suppose $\omega$ is a 1-form on $M$ and consider the trivial line bundle $E := M \times \mathbb{R} \to M$ with the connection $\nabla_X^\omega f := X \cdot f + \omega(X) f$. For its curvature one easily checks $R^{\nabla^\omega} = d\omega$. Via the obvious identification $\Omega^*(M; E) = \Omega^*(M)$ the covariant exterior derivative of $E$-valued differential forms is $d^{\nabla^\omega} = d^\omega$. For closed $\omega$ one has $d^\omega d^\omega = 0$, corresponding to the fact that the connection $\nabla^\omega$ is flat in this case. Moreover, for two 1-forms $\omega_1$ and

$\omega_2$ one has

$$d^{\omega_1+\omega_2}(\alpha \wedge \beta) = (d^{\omega_1}\alpha) \wedge \beta + (-1)^{|\alpha|}\alpha \wedge (d^{\omega_2}\beta), \tag{1}$$

which corresponds to the fact $\nabla^{\omega_1} \otimes \nabla^{\omega_2} = \nabla^{\omega_1+\omega_2}$.

*Remark 5.* Let $\omega$ be a closed 1-form on $M$. Since $d^\omega d^\omega = 0$ one obtains a 'twisted' deRham cohomology $H^*_{d^\omega}(M)$. It is easy to see, that this is the cohomology of the locally constant sheaf of $d^\omega$-constant functions, i.e. locally defined functions $f$, satisfying $d^\omega f = 0$. Note, that since $d^\omega\Omega = 0$, $\Omega$ defines a $d^\omega$-cohomology class $[\Omega] \in H^2_{d^\omega}(M)$, but even if $M$ is closed this class might vanish, cf. Example 3 below. From the derivation like formula (1) one sees, that the wedge product induces

$$\wedge : H^{k_1}_{d^{\omega_1}}(M) \times H^{k_2}_{d^{\omega_2}}(M) \to H^{k_1+k_2}_{d^{\omega_1+\omega_2}}(M).$$

If $\omega' = \omega - da$ for $a \in C^\infty(M,\mathbb{R})$ then $e^a : H^*_{d^\omega}(M) \cong H^*_{d^{\omega'}}(M)$. So for $[\omega] = 0 \in H^1(M)$ the 'twisted' deRham cohomology is isomorphic to the ordinary deRham cohomology. For $[\omega] \neq 0 \in H^1(M)$ one easily shows $H^0_{d^\omega}(M) = 0$ and $H^0_{d_c^\omega}(M) = 0$, where the latter denotes cohomology with compact supports. Together with the Poincaré duality

$$\mathrm{PD} : H^k_{d^\omega}(M) \to \left(H^{\dim(M)-k}_{d_c^{-\omega}}(M)\right)^*, \quad \mathrm{PD}(\alpha)(\beta) = \int_M \alpha \wedge \beta,$$

one obtains $H^{\dim(M)}_{d^\omega}(M) = 0$ and $H^{\dim(M)}_{d_c^\omega}(M) = 0$ for all $[\omega] \neq 0 \in H^1(M)$, see [9] and [10].

**Example 1.** Symplectic manifolds are l.c.s. manifolds with vanishing $\omega$. Moreover $(M,\Omega,\omega)$ is conformally equivalent to a symplectic manifold iff $[\omega] = 0 \in H^1(M)$.

**Example 2.** If $\omega$ is a closed 1-form on a manifold $N$ then $(T^*N, d^{\pi^*\omega}\theta, \pi^*\omega)$ is an l.c.s. manifold, where $\theta$ is the canonical 1-form on $T^*N$ and $\pi : T^*N \to N$ denotes the projection.

**Example 3 (Banyaga).** If $(N,\alpha)$ is a contact manifold then $(N \times S^1, d^\nu\alpha, \nu)$ is an l.c.s. manifold, where $\nu$ is the standard volume form on $S^1$. For example $S^3 \times S^1$ is an l.c.s. manifold, although it does not admit a symplectic structure, for $H^2(S^3 \times S^1) = 0$. Note, that $H^*_{d^\nu}(N \times S^1) = 0$, since one has a Künneth theorem and $H^*_{d^\nu}(S^1) = 0$.

**Example 4 (Guedira and Lichnerowicz).** There is a natural correspondence between even dimensional transitive Jacobi manifolds and l.c.s. manifolds. In particular, every even dimensional leave of a Jacobi manifold posses a natural l.c.s. structure, cf. [9]. The odd dimensional transitive Jacobi manifolds, are in natural one-to-one correspondence with contact structures. Particularly every odd dimensional leave of a Jacobi manifold is a contact manifold.

**Example 5 (Lee).** Suppose $\mathcal{U}$ is an open covering of $M$ and for every $U \in \mathcal{U}$, we have given an l.c.s. structure $(\Omega_U, \omega_U)$ on $U$, such that

$$(\Omega_U, \omega_U)|_{U \cap V} \sim (\Omega_V, \omega_V)|_{U \cap V},$$

for all $U, V \in \mathcal{U}$. One easily shows, cf. [16], that in this situation there exists an l.c.s. structure $(\Omega, \omega)$ on $M$, such that $(\Omega, \omega)|_U \sim (\Omega_U, \omega_U)$ for all $U \in \mathcal{U}$. Obviously $(\Omega, \omega)$ is unique up to conformal equivalence.

For an l.c.s. manifold $(M, \Omega, \omega)$ let us denote by

$$\mathrm{Diff}_c^\infty(M, \Omega, \omega) := \left\{ g \in \mathrm{Diff}_c^\infty(M) \mid (g^*\Omega, g^*\omega) \sim (\Omega, \omega) \right\}$$

the group of compactly supported diffeomorphisms preserving the conformal equivalence class of $(\Omega, \omega)$. The corresponding Lie algebra of vector fields is

$$\mathfrak{X}_c(M, \Omega, \omega) := \left\{ X \in \mathfrak{X}_c(M) \mid \exists c_X \in \mathbb{R} : L_X^\omega \Omega = c_X \Omega \right\},$$

where

$$L_X^\omega : \Omega^*(M) \to \Omega^*(M), \quad L_X^\omega \alpha := L_X \alpha + \omega(X)\alpha,$$

for a vector field $X \in \mathfrak{X}(M)$ and a closed 1-form $\omega$. Note that $\mathrm{Diff}_c^\infty(M, \Omega, \omega)$ and $\mathfrak{X}_c(M, \Omega, \omega)$ do only depend on the conformal equivalence class of $(M, \Omega, \omega)$.

*Remark 6.* Note, that for two closed 1-forms $\omega_1$ and $\omega_2$ one has

$$L_X^{\omega_1 + \omega_2}(\alpha \wedge \beta) = (L_X^{\omega_1} \alpha) \wedge \beta + \alpha \wedge (L_X^{\omega_2} \beta).$$

Moreover $L_X^\omega L_Y^\omega - L_Y^\omega L_X^\omega = L_{[X,Y]}^\omega$, $L_X^\omega d^\omega - d^\omega L_X^\omega = 0$, $L_X^\omega i_Y - i_Y L_X^\omega = i_{[X,Y]}$ and $d^\omega i_X + i_X d^\omega = L_X^\omega$, for a closed 1-form $\omega$ and vector fields $X, Y \in \mathfrak{X}(M)$.

**Theorem 1 (Weinstein chart).** *Suppose $(M, \Omega, \omega)$ is an l.c.s. manifold. Then $\mathrm{Diff}_c^\infty(M, \Omega, \omega)$ is a regular Lie group with Lie algebra $\mathfrak{X}_c(M, \Omega, \omega)$ in the sense of [15]. Particularly, if $M$ is compact then $\mathrm{Diff}^\infty(M, \Omega, \omega)$ is a regular Fréchet-Lie group with Lie algebra $\mathfrak{X}(M, \Omega, \omega)$.*

The proof, see [10] and [12], is a generalization of a well known construction due to A. Weinstein, cf. [24].

## 2 Infinitesimal invariants

Recall that $X \in \mathfrak{X}_c(M, \Omega, \omega)$ iff $L_X^\omega \Omega = c_X \Omega$ for some constant $c_X \in \mathbb{R}$. The following mapping, called extended Lee homomorphism cf. [9] or [23],

$$\varphi : \mathfrak{X}_c(M, \Omega, \omega) \to \mathbb{R}, \quad \varphi(X) := c_X$$

is a Lie algebra homomorphism, where $\mathbb{R}$ is considered as Abelian Lie algebra. Indeed, using the formulas in Remark 6 we get

$$
\begin{aligned}
L_{[X,Y]}^\omega \Omega &= L_X^\omega L_Y^\omega \Omega - L_Y^\omega L_X^\omega \Omega \\
&= L_X^\omega(c_Y \Omega) - L_Y^\omega(c_X \Omega) \\
&= (L_X c_Y)\Omega + c_Y L_X^\omega \Omega - (L_Y c_X)\Omega - c_X L_Y^\omega \Omega \\
&= (c_Y c_X - c_X c_Y)\Omega = 0.
\end{aligned}
$$

Note that $\mathfrak{X}_c(M,\Omega,\omega)$ and $\varphi$ do only depend on the conformal equivalence class of $(M,\Omega,\omega)$. Indeed, if $L_X^\omega \Omega = c_X \Omega$ and $a \in C^\infty(M,\mathbb{R})$, we get

$$
L_X^{\omega - da}(e^a \Omega) = (L_X^{-da} e^a)\Omega + e^a L_X^\omega \Omega = c_X(e^a \Omega),
$$

since obviously $L_X^{-da} e^a = 0$. Remark, that $\varphi$ has to vanish if $M$ is non-compact or $[\omega] = 0 \in H^1(M)$. So it does not appear in the symplectic case, at least as long as one considers compactly supported vector fields.

The next invariant generalizes the Flux homomorphism to the l.c.s. case. The concept is essentially due to E. Calabi, cf. [7]. The mapping

$$
\psi : \ker \varphi \to H_{d_c^\omega}^1(M), \quad \psi(X) := [i_X \Omega]
$$

is a surjective Lie algebra homomorphism, where $H_{d_c^\omega}^1(M)$ is considered as Abelian Lie algebra. It is well defined and onto, since $X \in \mathfrak{X}_c(M,\Omega,\omega)$ iff $d^\omega i_X \Omega = L_X^\omega \Omega = 0$ and since $\Omega$ is non-degenerate. To see that $\psi$ vanishes on brackets one uses the formulas in Remark 6 to get

$$
\begin{aligned}
i_{[X,Y]}\Omega &= L_X^\omega i_Y \Omega - i_Y L_X^\omega \Omega \\
&= d^\omega i_X i_Y \Omega + i_X d^\omega i_Y \Omega \\
&= d^\omega i_X i_Y \Omega + i_X L_Y^\omega \Omega - i_X i_Y d^\omega \Omega = d^\omega i_X i_Y \Omega.
\end{aligned}
$$

If $\omega' = \omega - da$ then $e^a : H_{d_c^\omega}^*(M) \cong H_{d_c^{\omega'}}^*(M)$, see Remark 5, and $\psi'(X) = e^a \psi(X)$. In particular, $\ker \psi$ does only depend on the conformal equivalence class of $(M,\Omega,\omega)$.

Suppose $M$ is of dimension $2n$. Then the following is a surjective Lie algebra homomorphism

$$
\rho : \ker \psi \to H_{d_c^\omega}^{2n}(M)\big/\big(H_{d_c^\omega}^0(M) \cdot [\Omega^n]\big), \quad \rho(X) := [h_X \Omega^n],
$$

where $h_X$ is a function with compact support, such that $d^\omega h_X = i_X \Omega$. Again $\ker \rho$ does only depend on the conformal equivalence class of $(M,\Omega,\omega)$. But note, that the codomain of $\rho$ is non-trivial only if $M$ is non-compact and $[\omega] = 0 \in H^1(M)$, i.e. $(M,\Omega,\omega)$ is conformally equivalent to a symplectic manifold. The invariant $\rho$ is sometimes called Calabi invariant, cf. [7], [2], [10] and [11].

*Remark 7.* Note, that $X \in \ker \rho$ iff there exist $h \in \Omega_c^0(M)$ and $\alpha \in \Omega_c^{2n-1}(M)$, such that $i_X\Omega = d^w h$ and $h\Omega^n = d^{(n+1)w}\alpha$.

The following is due to E. Calabi in the symplectic case.

**Theorem 2.** *Let* $(M, \Omega, \omega)$ *be an l.c.s. manifold. Then* $\ker \rho$ *is a perfect Lie algebra, i.e.* $[\ker \rho, \ker \rho] = \ker \rho$.

*Remark 8.* Theorem 2 permits to compute the derived series[a] of the Lie algebra $\mathfrak{X}_c(M, \Omega, \omega)$, see [10]. The only algebras in the derived series of $\mathfrak{X}_c(M, \Omega, \omega)$ are $\ker \varphi$, $\ker \psi$ and $\ker \rho$, but there is a case[b], where $[\mathfrak{X}_c(M, \Omega, \omega), \mathfrak{X}_c(M, \Omega, \omega)] = \ker \rho$ and $\ker \psi \neq \ker \rho$. However, in any case one has $D^2 \mathfrak{X}_c(M, \Omega, \omega) = \ker \rho$.

If $(M, \Omega, \omega)$ is an l.c.s. manifold and $U \subseteq M$ open, we will denote by $\varphi_U$, $\psi_U$ and $\rho_U$ the corresponding invariants of the l.c.s. manifold $(U, \Omega|_U, \omega|_U)$. Note that we have $\mathfrak{X}_c(U, \Omega|_U, \omega|_U) \subseteq \mathfrak{X}_c(M, \Omega, \omega)$, $\ker \varphi_U \subseteq \ker \varphi$, $\ker \psi_U \subseteq \ker \psi$ and $\ker \rho_U \subseteq \ker \rho$, given by extending a vector field by zero.

**Lemma 1 (Fragmentation lemma).** *Let* $(M, \Omega, \omega)$ *be an l.c.s. manifold,* $\mathcal{U}$ *be an open covering of $M$ and suppose* $X \in \mathfrak{X}_c(M, \Omega, \omega)$. *Then for every* $X \in \ker \rho$ *there exist* $N \in \mathbb{N}$, $U_1, \ldots, U_N \in \mathcal{U}$ *and* $X_i \in \ker \rho_{U_i}$, *such that* $X = \sum_{i=1}^N X_i$.

*Proof.* By Remark 7 we find $h \in \Omega_c^0(M)$ and $\alpha \in \Omega_c^{2n-1}(M)$, such that $i_X\Omega = d^w h$ and $h\Omega^n = d^{(n+1)w}\alpha$. Choose $N \in \mathbb{N}$ and $U_1, \ldots, U_N \in \mathcal{U}$ which cover $\operatorname{supp}\alpha$. Choose a partition of unity $\{\lambda_0, \lambda_1, \ldots, \lambda_N\}$ subordinated to the open covering $\{M \setminus \operatorname{supp}\alpha, U_1, \ldots, U_N\}$, i.e. $\operatorname{supp}\lambda_0 \subseteq M \setminus \operatorname{supp}\alpha$ and $\operatorname{supp}\lambda_i \subseteq U_i$ for $1 \leq i \leq N$. For $0 \leq i \leq N$ define $h_i \in \Omega_c^0(M)$, by $h_i\Omega^n := d^{(n+1)w}(\lambda_i\alpha)$ and $X_i \in \mathfrak{X}_c(M, \Omega, \omega)$ by $i_{X_i}\Omega := d^w h_i$. Since we have

$$\sum_{i=0}^N h_i\Omega^n = d^{(n+1)w}\sum_{i=0}^N \lambda_i\alpha = d^{(n+1)w}\alpha = h\Omega^n,$$

we get $\sum_{i=0}^N h_i = h$ and hence $\sum_{i=0}^N X_i = X$. Moreover $X_0 = 0$ and $X_i \in \ker \rho_{U_i}$ for $1 \leq i \leq N$. $\qquad\square$

The following can be found in [1], where it is used to show Theorem 2 in the symplectic case.

**Lemma 2.** *Let* $U \subseteq V$ *be open subsets of* $\mathbb{R}^{2n}$ *equipped with the standard symplectic form* $\Omega = dx^1 \wedge dx^2 + \cdots$, *such that* $\bar{U} \subseteq V$. *For all* $X \in \ker \rho_U$

---

[a] The derived series of a Lie algebra $\mathfrak{g}$ is defined by $D^0\mathfrak{g} := \mathfrak{g}$ and $D^{i+1}\mathfrak{g} := [D^i\mathfrak{g}, D^i\mathfrak{g}]$.

[b] $M$ non-compact, $[\omega] = 0 \in H^1(M)$ and vanishing symplectic pairing

$$\{\cdot, \cdot\} : H^1_{d_c^w}(M) \times H^1_{d_c^w}(M) \to H^{2n}_{d_c^{(n+1)w}}(M), \quad \{\alpha_1, \alpha_2\} := \alpha_1 \wedge \alpha_2 \wedge [\Omega^{n-1}].$$

*there exist* $Y_i, Z_i \in \ker \rho_V$, *such that* $X = \sum_{i=1}^{2n} [Y_i, Z_i]$. *Particularly* $\ker \rho_U \subseteq [\ker \rho_V, \ker \rho_V]$.

*Proof of Theorem 2.* We have to show $\ker \rho \subseteq [\ker \rho, \ker \rho]$. In view of Lemma 1 we may assume, that $(M, \Omega, \omega)$ is an open ball in $\mathbb{R}^{2n}$ equipped with the standard symplectic structure, but this case follows from Lemma 2. $\square$

A well known theorem of L.E. Pursell and M.E. Shanks see [19] states, roughly speaking, that a smooth manifold is completely determined by its Lie algebra of vector fields. More precisely, if there exists an isomorphism of the Lie algebras of vector fields then there exists a unique diffeomorphism between the manifolds, inducing the given Lie algebra isomorphism. H. Omori proved several generalizations, namely the Lie algebra of vector fields preserving a symplectic form resp. a volume form uniquely determines the manifold together with the symplectic resp. volume structure up to multiplication with a constant, see [18]. We will show an analogous statement for l.c.s. structures, i.e. any of the Lie algebras $\mathfrak{X}_c(M, \Omega, \omega)$, $\ker \varphi$, $\ker \psi$, $\ker \rho$ uniquely determines the l.c.s. manifold $(M, \Omega, \omega)$ up to conformal equivalence. More precisely one has the following

**Theorem 3.** *Let* $(M_i, \Omega_i, \omega_i)$, $i = 1, 2$, *be two l.c.s. manifolds and assume that* $\kappa$ *is a Lie algebra isomorphism from one of the Lie algebras* $\mathfrak{X}_c(M_1, \Omega_1, \omega_1)$, $\ker \varphi_1$, $\ker \psi_1$, $\ker \rho_1$ *onto one of the Lie algebras* $\mathfrak{X}_c(M_2, \Omega_2, \omega_2)$, $\ker \varphi_2$, $\ker \psi_2$, $\ker \rho_2$. *Then there exists a unique diffeomorphism* $g : M_1 \to M_2$, *such that* $\kappa = g_*$. *Moreover we have* $(M_1, \Omega_1, \omega_1) \sim (M_1, g^* \Omega_2, g^* \omega_2)$.

For the proof we first state a slightly weaker statement and a simple lemma.

**Proposition 1.** *Let* $(M_i, \Omega_i, \omega_i)$, $i = 1, 2$, *be two l.c.s. manifolds and let* $\kappa : \ker \rho_1 \to \ker \rho_2$ *a Lie algebra isomorphism. Then there exists a unique diffeomorphism* $g : M_1 \to M_2$ *such that* $\kappa = g_*$. *Moreover* $(M_1, \Omega_1, \omega_1) \sim (M_1, g^* \Omega_2, g^* \omega_2)$.

**Lemma 3.** *Let* $\mathfrak{g}$ *be a Lie algebra such that* $\mathrm{ad} : \mathfrak{g} \to L([\mathfrak{g}, \mathfrak{g}])$ *is injective and let* $\lambda : \mathfrak{g} \to \mathfrak{g}$ *be a Lie algebra homomorphism, such that* $\lambda|_{[\mathfrak{g},\mathfrak{g}]} = \mathrm{id}$. *Then* $\lambda = \mathrm{id}$.

*Proof.* For $X \in \mathfrak{g}$ and $Y \in [\mathfrak{g}, \mathfrak{g}]$ we have $[X, Y] = \lambda([X, Y]) = [\lambda(X), \lambda(Y)] = [\lambda(X), Y]$, hence $\mathrm{ad}(X - \lambda(X)) = 0 \in L([\mathfrak{g}, \mathfrak{g}])$ and so $\lambda(X) = X$. $\square$

*Proof of Theorem 3.* We have a Lie algebra isomorphism $\kappa : \mathfrak{g}_1 \to \mathfrak{g}_2$. Since $D^2 \mathfrak{g}_i = \ker \rho_i$, see Remark 8, $\kappa$ restricts to an isomorphism $\kappa|_{\ker \rho_1} : \ker \rho_1 \to \ker \rho_2$. From Proposition 1 we obtain a unique diffeomorphism $g : M_1 \to M_2$ such that $\kappa|_{\ker \rho_1} = g_*|_{\ker \rho_1}$. Moreover $(M_1, \Omega_1, \omega_1) \sim (M_1, g^* \Omega_2, g^* \omega_2)$. So

$g^*\mathfrak{g}_2$ is one of the Lie algebras $\mathfrak{X}_c(M,\Omega_1,\omega_1)$, $\ker\varphi_1$, $\ker\psi_1$, $\ker\rho_1$ and we either have $g^*\mathfrak{g}_2 \subseteq \mathfrak{g}_1$ or $g^*\mathfrak{g}_2 \supseteq \mathfrak{g}_1$. Assume we are in the first case (for the second consider $g^{-1}$). Then $\lambda := g_*^{-1}\circ\kappa = g^*\circ\kappa : \mathfrak{g}_1 \to \mathfrak{g}_1$ is a Lie algebra homomorphism and we know that $\lambda|_{D^2\mathfrak{g}_1} = \mathrm{id}$. Moreover we obviously have for every vector field $Z \in \mathfrak{X}(M_1)$ the following property:

$$[Z, X] = 0 \quad \text{for all } X \in \ker\rho_1 \quad \Rightarrow \quad Z = 0.$$

Using $\ker\rho_1 \subseteq D^{i+1}\mathfrak{g}_1$ we obtain $\mathrm{ad} : D^i\mathfrak{g}_1 \to L([D^i\mathfrak{g}_1, D^i\mathfrak{g}_1]) = L(D^{i+1}\mathfrak{g}_1)$ is injective for all $i$. So we can apply Lemma 3 twice to obtain $\lambda|_{D^1\mathfrak{g}_1} = \mathrm{id}$ and $\lambda = \lambda|_{D^0\mathfrak{g}_1} = \mathrm{id}$, i.e. $g_* = \kappa$. $\qquad\square$

*Proof of Proposition 1.* We will only sketch the proof, since it is very similar to the symplectic case, as soon as one has the fragmentation Lemma 1. For more details see [10] or [18] in the symplectic case.

Suppose $(M,\Omega,\omega)$ is an l.c.s. manifold. Then for every $x \in M$,

$$I_x := \{X \in \ker\rho \mid X \text{ is flat at } x\}$$

is a maximal ideal in $\ker\rho$, and every maximal ideal of $\ker\rho$ is of this form. This permits to define a bijection $g : M_1 \to M_2$ by $I_{g(x)} = \kappa(I_x)$. For every $A \subseteq M$ one has

$$\bar{A} = \Big\{x \in M \mid \bigcap_{y\in A} I_y \subseteq I_x\Big\},$$

and hence $g$ is a homeomorphism. Next one shows, that for $X \in \ker\rho$ and $x \in M$

$$X(x) \neq 0 \quad \Leftrightarrow \quad [X, \ker\rho] + I_x = \ker\rho.$$

So $X_i \in \ker\rho_1$ are linearly independent at $x$ iff $\kappa(X_i)$ are linearly independent at $g(x)$. Using a Darboux chart and vector fields in $\ker\rho_1$ involving the coordinate functions, one shows, that $g$ is a diffeomorphism and that $(g^*\Omega_2, g^*\omega_2) \sim (\Omega_1, \omega_1)$. $\qquad\square$

## 3 Integrated invariants

A well known theorem of W.P. Thurston states that $\mathrm{Diff}_c^\infty(M)_o$ is a simple group, see [22]. His proof used a theorem due to M.R. Herman see [14], which solves the problem for the torus. The group of volume preserving diffeomorphisms is not simple in general, but there exists a homomorphism and its kernel is simple. This was shown by W.P. Thurston, see [6]. A. Banyaga showed an analogous statement in the symplectic case, see [2]. In [11] this was generalized to l.c.s. manifolds.

The infinitesimal invariants $\varphi$, $\psi$ and $\rho$ can by integrated to group homomorphisms $\Phi$, $\Psi$ and $R$, see [11]. The kernels of these homomorphisms do only depend on the conformal equivalence class of $(M, \Omega, \omega)$.

**Theorem 4.** *Let $(M, \Omega, \omega)$ be an l.c.s. manifold. Then $\ker R$ is a simple group.*

The main ingredients in the proof are Banyaga's theorem for symplectic open balls and the following fragmentation lemma, which can be found in [10] or [11].

**Lemma 4 (Fragmentation lemma).** *Let $(M, \Omega, \omega)$ be an l.c.s. manifold and let $\mathcal{U}$ be an open covering of $M$. Then for any $g \in C^\infty\big((I, 0), (\ker R, \mathrm{id})\big)$ there exist $N \in \mathbb{N}$, $U_i \in \mathcal{U}$ and $g^i \in C^\infty\big((I, 0), (\ker R_{U_i}, \mathrm{id})\big)$, such that $g_t = g_t^1 \circ \cdots \circ g_t^N$, for all $t \in I = [-1, 1]$.[c]*

*Remark 9.* Theorem 4 permits to compute the derived series[d] of the connected component $\mathrm{Diff}_c^\infty(M, \Omega, \omega)_o$. All groups in this derived series are $\ker \Phi$, $\ker \Psi$ or $\ker R$ and it is precisely the integral counterpart of the derived series of $\mathfrak{X}_c(M, \Omega, \omega)$, cf. Remark 8. Particularly one obtains $D^2 \mathrm{Diff}_c^\infty(M, \Omega, \omega)_o = \ker R$ for every l.c.s. manifold $(M, \Omega, \omega)$.

Filipkiewicz showed that a smooth manifold is uniquely determined by its group of diffeomorphisms. That is, if two manifold have isomorphic diffeomorphism groups then the underlying manifolds are diffeomorphic, see [8]. He used techniques developed in [25] and [21] who proved analogous statements in the topological setting. There are many generalizations to other geometric structures, see [3], [4], [5] and [20]. A similar result is true for l.c.s. manifolds. More precisely we have

**Theorem 5.** *Let $(M_i, \Omega_i, \omega_i)$ be two l.c.s. manifolds, $i = 1, 2$, and suppose $\kappa : G_1 \to G_2$ is an isomorphism from one of the groups $\mathrm{Diff}_c^\infty(M_1, \Omega_1, \omega_1)_o$, $\ker \Phi_1$, $\ker \Psi_1$, $\ker R_1$ onto one of the groups $\mathrm{Diff}_c^\infty(M_2, \Omega_2, \omega_2)_o$, $\ker \Phi_2$, $\ker \Psi_2$, $\ker R_2$. Then there exists a unique homeomorphism $g : M_1 \to M_2$ such that $\kappa(h) = g \circ h \circ g^{-1}$ for all $h \in G_1$. Moreover $g$ is a diffeomorphism and $(M_1, \Omega_1, \omega_1) \sim (M_1, g^*\Omega_2, g^*\omega_2)$.*

To prove Theorem 5 one first observes, that since one has the fragmentation Lemma 4 and Proposition 1, a result due to T. Rybicki, see [20], can be applied, which yields the following

**Proposition 2.** *Let $(M_i, \Omega_i, \omega_i)$, $i = 1, 2$, be two l.c.s. manifolds and suppose $\kappa : \ker R_1 \to \ker R_2$ is an isomorphism of groups. Then there exists a unique*

---

[c]For an l.c.s. manifold $(M, \Omega, \omega)$ and $U \subseteq M$ open, $R_U$ denotes the corresponding invariant for the l.c.s. manifold $(U, \Omega|_U, \omega|_U)$. In particular every $g^i$ in the proposition is supported in $U_i$.

[d]Recall, that for a group $G$, the derived series is defined by $D^0 G := G$ and $D^{i+1} G := [D^i G, D^i G]$.

*homeomorphism $g : M_1 \to M_2$ such that $\kappa(h) = g \circ h \circ g^{-1}$ for all $h \in \ker R_1$. Moreover $g$ is a diffeomorphism and $(M_1, \Omega_1, \omega_1) \sim (M_1, g^*\Omega_2, g^*\omega_2)$.*

Moreover, we have a group analogue to Lemma 3:

**Lemma 5.** *Let $G$ be a group such that $\operatorname{conj} : G \to \operatorname{Aut}([G,G])$ is injective and let $\lambda : G \to G$ be a homomorphism, such that $\lambda|_{[G,G]} = \operatorname{id}$. Then $\lambda = \operatorname{id}$.*

*Proof.* For $g \in G$ and $h \in [G,G]$ we have $[g,h] = \lambda([g,h]) = [\lambda(g), \lambda(h)] = [\lambda(g), h]$, hence $\operatorname{conj}_{g^{-1}\lambda(g)} = \operatorname{id} \in \operatorname{Aut}([G,G])$, and by injectivity $\lambda(g) = g$. $\qquad\square$

*Proof of Theorem 5.* The restriction of $\kappa$ is an isomorphism $\kappa|_{D^2G_1} : D^2G_1 \to D^2G_2$. In any case $D^2G_i = \ker R_i$ for $i = 1, 2$, by Remark 9. So we may apply Proposition 2 and obtain a unique homeomorphism $g : M_1 \to M_2$ such that $\kappa(h) = ghg^{-1}$ for all $h \in \ker R_1 = D^2G_1$. Moreover $g$ is a diffeomorphism and $(M_1, \Omega_1, \omega_2) \sim (M_1, g^*\Omega_2, g^*\omega_2)$. So it remains to show that $\kappa(h) = \operatorname{conj}_g(h) := ghg^{-1}$ for all $h \in G_1$. From $(M_1, \Omega_1, \omega_2) \sim (M_1, g^*\Omega_2, g^*\omega_2)$ we see that $\operatorname{conj}_{g^{-1}}(G_2) \subseteq G_1$ or $\operatorname{conj}_{g^{-1}}(G_2) \supseteq G_1$. Assume we are in the first case (otherwise consider $g^{-1}$). Then $\lambda := \operatorname{conj}_{g^{-1}} \circ \kappa : G_1 \to G_1$ is a homomorphism and $\lambda|_{D^2G_1} = \operatorname{id}$. Moreover, for $h \in \operatorname{Diff}^\infty(M_1)$ we have:

$$[h, k] = \operatorname{id} \quad \forall k \in \ker R_1 \quad \Rightarrow \quad h = \operatorname{id},$$

since $\ker R_1$ acts 2-transitive on $M_1$. Using $\ker R_1 \subseteq D^{i+1}G_1$ we obtain $\operatorname{conj} : D^iG_1 \to \operatorname{Aut}([D^iG_1, D^iG_1]) = \operatorname{Aut}(D^{i+1}G_1)$ is injective for all $i$. So we can apply Lemma 5 twice and obtain first $\lambda|_{D^1G_1} = \operatorname{id}$ and finally $\lambda = \lambda_{D^0G_1} = \operatorname{id}$, i.e. $\kappa = (\operatorname{conj}_g)|_{G_1}$. $\qquad\square$

## 4  Symplectic reduction

The aim of this last section is to show that, in the regular case, the symplectic reduction is possible for l.c.s. manifolds. In the paper of Marsden and Weinstein [17] it has been formalized the fact that if an $n$-dimensional symmetry group acts on a Hamiltonian system then the number of degrees of freedom can be reduced by $n$, and the dimension of the phase space is reduced by $2n$. This is still true for l.c.s. manifolds. For proofs see [13].

Every l.c.s. manifold is a Jacobi manifold, so $C^\infty(M, \mathbb{R})$ has a Lie algebra structure. More explicitly, the bracket is given by

$$\{f_1, f_2\} := \Omega(X_{f_1}, X_{f_2}) = L^\omega_{X_{f_2}} f_1 = -L^\omega_{X_{f_1}} f_2.$$

Here $X_f$ denotes the Hamiltonian vector field to $f \in C^\infty(M, \mathbb{R})$ given by $i_{X_f}\Omega = d^\omega f$. One has $X_{\{f_1, f_2\}} = -[X_{f_1}, X_{f_2}]$ and

$$0 \to H^0_{d^\omega}(M) \to C^\infty(M; R) \to \operatorname{Ham}(M, \Omega, \omega) \to 0$$

is, up to a sign, a central extension of Lie algebras, where $\mathrm{Ham}(M, \Omega, \omega)$ denotes the Lie algebra of Hamiltonian vector fields. For compact $M$ one has $\mathrm{Ham}(M, \Omega, \omega) = \ker \psi$, but note that in this section we consider vector fields with arbitrary support.

Let $G$ be a finite dimensional Lie group with Lie algebra $\mathfrak{g}$, acting from the left on $M$. We will write $l_g : M \to M$ for the action of the element $g \in G$. We will always assume that $G$ acts symplectically, i.e. $l_g \in \mathrm{Diff}^\infty(M, \Omega, \omega)$ for all $g \in G$. For $Y \in \mathfrak{g}$ let $\zeta_Y \in \mathfrak{X}(M, \Omega, \omega)$ denote the fundamental vector field to $Y$. If there exists a Lie algebra homomorphism $\hat{\mu} : \mathfrak{g} \to C^\infty(M, \mathbb{R})$, such that $X_{\hat{\mu}(Y)} = \zeta_Y$ for all $Y \in \mathfrak{g}$, we define $\mu : M \to \mathfrak{g}^*$ by $\langle \mu(x), Y \rangle = \hat{\mu}(Y)(x)$, $Y \in \mathfrak{g}$, $x \in M$, and call $\mu$ an equivariant moment mapping for the action. Note, that in this case the action of $G_o$ is Hamiltonian. One easily shows

**Proposition 3.** *If $H^2(\mathfrak{g}; H^0_{d^\omega}(M)) = 0$, where $H^0_{d^\omega}(M)$ is considered as trivial $\mathfrak{g}$-module, then there exists an equivariant moment mapping. Moreover, if an equivariant moment mapping exists, then the set of all equivariant moment mappings is naturally parametrized by $H^1(\mathfrak{g}; H^0_{d^\omega}(M))$.*

*Remark* 10. If $\mathfrak{g}$ is semi simple then there always exists a unique equivariant moment mapping, since $H^2(\mathfrak{g}; H^0_{d^\omega}(M)) = 0$ and $H^1(\mathfrak{g}; H^0_{d^\omega}(M)) = 0$ by the second and the first Whitehead lemma, respectively.

If the l.c.s. manifold is not conformally equivalent to a symplectic manifold, i.e. $[\omega] \neq 0 \in H^1(M)$, then there always exists a unique equivariant moment mapping, for $H^0_{d^\omega}(M) = 0$ in this case.

Suppose that the l.c.s. structure is exact, i.e. $\Omega = d^\omega \theta$, and that the action preserves $\theta$, i.e. $L^\omega_{\zeta_Y} \theta = 0$. Then $\hat{\mu}(Y) := -i_{\zeta_Y} \theta$ defines an equivariant moment mapping.

If the $G$-action admits an equivariant moment mapping $\mu$, then the vector fields which are $\Omega$-orthogonal to the orbits span an involutive distribution. Suppose $L$ is a maximal connected submanifold of $M$, tangential to this distribution. Then $\hat{\mu}(L) \subseteq \mathbb{R}^+ \cdot \mu(x_0)$ for every $x_0 \in L$. However $\hat{\mu}$ need not be constant along $L$. Let $\mathfrak{g}_L := \{Y \in \mathfrak{g} : \mathrm{ad}^*_Y \hat{\mu}(x_0) = 0\}$, which does not depend on $x_0 \in \mathfrak{g}$. Then $\mathfrak{g}_L$ are precisely those $Y \in \mathfrak{g}$, for which $\zeta_Y$ is tangential to $L$. Since $L$ is maximal, the connected subgroup of $G$ corresponding to $\mathfrak{g}_L \subseteq \mathfrak{g}$ leaves $L$ invariant.

In [13] the following generalization of symplectic reduction, see [17], is proved:

**Theorem 6.** *Let $G$ be a finite dimensional Lie group acting symplectically on an l.c.s. manifold $(M, \Omega, \omega)$ and assume that the action admits an equivariant moment mapping $\mu$. Suppose $L$ is a maximal connected submanifold of $M$ with $T_x L = \zeta_{\mathfrak{g}}(x)^\perp$. Let $G_L$ be a subgroup of $G$ which preserves $L$ and has $\mathfrak{g}_L$ as Lie algebra, and assume that $G_L$ acts freely and properly on $L$. Then $P_L := L/G_L$ admits a unique (up to conformal equivalence) l.c.s. structure*

$(\bar{\Omega},\bar{\omega})$, such that $(L,i^*\Omega,i^*\omega) \sim (L,\pi^*\bar{\Omega},\pi^*\bar{\omega})$, where $i:L\to M$ denotes the inclusion and $\pi:L\to P_L$ denotes the projection.

**Remark 11.** Suppose $(M,\Omega,\omega)$ is an l.c.s. manifold with symplectic $G$-action and equivariant moment mapping $\mu$. If $a\in C^\infty(M,\mathbb{R})$ and $(\Omega',\omega') = (e^a\Omega,\omega-da)$, then $e^a\mu$ is an equivariant moment mapping for the same action on $(M,\Omega',\omega')$. Moreover $L$ is $\Omega$-orthogonal to the orbits iff it is $\Omega'$-orthogonal to them, and in this case $\mathfrak{g}_L = \mathfrak{g}'_L$. So the reduced space does only depend on the conformal equivalence class of $(M,\Omega,\omega)$.

**Example 6.** Let $G$ be a discrete group acting freely, properly and symplectic on an l.c.s. manifold $(M,\Omega,\omega)$. Then $\mathfrak{g}=0$, $\mu=0$ is an equivariant moment mapping, and the only possible choice for $L$ is $L=M$. If we choose $G_L = G$ then $P_L = M/G$ carries an l.c.s. structure. Notice that even if we start with a symplectic manifold, $M/G$ need not be conformally equivalent to a symplectic manifold.

**Example 7.** Suppose $(N,\alpha)$ is a contact manifold, such that the Reeb vector field $E$ generates a free $S^1$-action. Then $N/S^1$ inherits a symplectic structure. On the other hand the l.c.s. manifold $(N\times S^1, d^\nu\alpha, \nu)$, see Example 3, has an $S^1$-action generated by $E\times 0$. This action is Hamiltonian with equivariant moment mapping $\mu=$ const. Then $L=N\times\{\text{point}\}$, $\mathfrak{g}_L=\mathbb{R}$ and $G_L=S^1$ satisfy the assumptions of Theorem 6, and $L/G_L \cong N/S^1$ as l.c.s. manifolds.

**Remark 12.** Moreover one can show, that a Hamiltonian system which is invariant under the action, descends to a Hamiltonian system on the reduced space, see [13].

## Acknowledgments

The author is supported by the 'Fonds zur Förderung der wissenschaftlichen Forschung' (Austrian Science Fund), project number P14195-MAT.

## References

1. A. Avez, A. Lichnerowicz and A. Diaz-Miranda, *Sur l'algèbre des auto-morphismes infinitésimaux d'une variété symplectique*, J. Diff. Geom. **9** (1974), 1–40.
2. A. Banyaga, *Sur la structure du groupe des difféomorphismes qui préservent une forme symplectique*, Comm. Math. Helv. **53** (1978), 174–227.
3. A. Banyaga, *On isomorphic classical diffeomorphism groups. I*, Proc. Amer. Math. Soc. **98** (1986), 113–118.
4. A. Banyaga, *On isomorphic classical diffeomorphism groups. II*, J. Diff. Geom. **28** (1988), 23–35.

5. A. Banyaga and A. McInerney, *On isomorphic classical diffeomorphism groups. III*, Ann. Global Anal. Geom. **13** (1995), 117–127.

6. A. Banyaga, *The Structure of Classical Diffeomorphism Groups*, Mathematics and its Applications, Kluwer Acad. Publ., 1997.

7. E. Calabi, *On the group of automorphisms of a symplectic manifold*, in Problems in Analysis, Symposium in Honor of S. Bochner, 1969, Princeton University Press, Princeton, NJ, 1970, 1–26.

8. R.P. Filipkiewicz, *Isomorphisms between diffeomorphism groups*, Ergodic Theory Dyn. Sys. **2** (1982), 159–171.

9. F. Guedira and A. Lichnerowicz, *Géometrie des algèbres de Lie locales de Kirillov*, J. Math. Pures Appl. **63** (1984), 407–484.

10. S. Haller, *Perfectness and Simplicity of Certain Groups of Diffeomorphisms*, Thesis, University of Vienna, 1998.

11. S. Haller and T. Rybicki, *On the group of diffeomorphisms preserving a locally conformal symplectic structure*, Ann. Global Anal. Geom. **17** (1999), 475–502.

12. S. Haller and T. Rybicki, *Integrability of the Poisson algebra on a locally conformal symplectic manifold*, The proceedings of the 19th winter school 'Geometry and Physics' (Srní, 1999) Rend. Circ. Mat. Palermo (II) Suppl. **63** (2000), 89–96.

13. S. Haller and T. Rybicki, *Reduction for locally conformal symplectic manifolds*, J. Geom. Phys. **37** (2001), 262–271.

14. M.R. Herman, *Sur le groupe des difféomorphismes du tore*, Ann. Inst. Fourier **23** (1973), 75–86.

15. A. Kriegl and P.W. Michor, *The Convenient Setting of Global Analysis*, Mathematical Surveys and Monographs **53**, Amer. Math. Soc., 1997.

16. H.-C. Lee, *A kind of even-dimensional differential geometry and its application to exterior calculus*, Amer. J. Math. **65** (1943), 433–438.

17. J. Marsden and A. Weinstein, *Reduction of symplectic manifolds with symmetry*, Rep. Math. Phys. **5** (1974), 121–130.

18. H. Omori, *Infinite Dimensional Lie Transformation Groups*, Lecture Notes in Mathematics, Springer Verlag **427**, 1974.

19. L.E. Pursell and M.E. Shanks, *The Lie algebra of smooth manifolds*, Proc. Amer. Math. Soc. **5** (1954), 468–472.

20. T. Rybicki, *Isomorphisms between groups of diffeomorphisms*, Proc. Amer. Math. Soc. **123** (1995), 303–310.

21. F. Takens, *Characterization of a differentiable structure by its group of diffeomorphisms*, Bol. Soc. Bras. Mat. **10** (1979), 17–25.

22. W.P. Thursten, *Foliations and groups of diffeomorphisms*, Bull. Amer. Math. Soc. **80** (1974), 304–307.

23. I. Vaisman, *Locally conformal symplectic manifolds*, Int. J. Math. Math. Sci. **8** (1985), 521–536.
24. A. Weinstein, *Symplectic manifolds and their Lagrangian submanifolds*, Adv. Math. **6** (1971), 329–346.
25. J.V. Whittaker, *On isomorphic groups and homeomorphic spaces*, Ann. Math. **78** (1963), 74–91.

# CRITICALITY OF UNIT CONTACT VECTOR FIELDS

PHILIPPE RUKIMBIRA

*Department of Mathematics, Florida International University, Miami FL 33199, USA*

*E-mail: rukim@fiu.edu*

This paper contains a characterization of K-contact unit vector fields in terms of J-holomorphic embeddings into the tangent unit sphere bundle. A consequence of this characterization is that these K-contact unit vector fields are minimal and harmonic sections. It is also proved that any closed flat contact manifold admits a parallelization by three critical unit vector fields, one parallel (hence minimizing), the other two are contact, non-Killing but minimize neither the volume nor the energy functionals.

## 1  Introduction

This paper is an extended version of the talk the author gave at the *International Conference on Infinite Dimensional Lie Groups in Geometry and Representation Theory* at Howard University, Washington DC in August 2000. Part of the extension constituted the talk given by the author at the conference *Foliations and Geometry 2001* held at Pontificia Universitade Catolica in Rio De Janeiro, Brazil in August 2001.

The volume and energy functionals are defined on the space of unit vector fields on a riemannian manifold $M$. Looking at Reeb vector fields as maps from contact metric manifolds to their tangent unit sphere bundles endowed with the usual Sasaki metric, we prove the following:

**Theorem A** *The Reeb vector field of a contact form $\alpha$ is a harmonic map if and only if $\alpha$ is a K-contact form (that is, the Reeb vector field is Killing).*

¿From an isometric embedding point of view, that is, considering the induced metric instead of the original contact metric on $M$, one has:

**Theorem A'** *The Reeb vector field of a contact form $\alpha$ is a minimal embedding if and only if $\alpha$ is a K-contact form.*

A modified energy functional will also be considered and it will be shown here that Reeb vector fields of sasakian-Einstein structures are its absolute minimizers. In dimension 3, the modified functional coincides with the original energy functional up to normalization.

The paper concludes with some examples of non-Killing critical unit vector fields arising from the flat contact metric geometry. The author would like to point out that reference [7] contains a unified study of the energy and the volume functionals. As it turns out, each of the two functionals is the

restriction of a general functional defined on the cross product of the space of
all unit vector fields with that of all riemannian metrics on $M$.

## 2    The volume functional

Let us denote by $\mathcal{X}^1(M)$ the space of unit vector fields on a closed riemannian
manifold $(M, g)$ and by $T^1M$ the tangent unit sphere bundle. We will denote
by $\pi_*: TTM \to TM$ and $\kappa: TTM \to TM$ the projection and the Levi-Civita
connection maps respectively. The volume $\mathcal{F}(V)$ of an element $V \in \mathcal{X}^1(M)$
is defined to be the volume of the submanifold $V(M) \subset T^1M$ where $T^1M$ is
endowed with the restriction of the usual Sasaki metric $g_s$ (see reference [?]).
We recall that $g_s$ is defined by:

$$g_s(X, Y) = g(\pi_* X, \pi_* Y) + g(\kappa X, \kappa Y).$$

The metric $g_v$ induced on $M$ by the map $V: M \to T^1M$ is related to the
metric $g$ on $M$ by the identity (see [?], [?])

$$g_v(X, Y) = g(\mathcal{L}_V X, Y),$$

where $\mathcal{L}_V = Id + (\nabla V)^t \circ \nabla V$ is a $g$-symmetric $(1,1)$ tensor field ($(\nabla V)^t$ stands
for the transpose of the operator $\nabla V$).

The problem of determining the condition for an element of $\mathcal{X}^1(M)$ to
be a critical point of the volume functional was tackled in references [?] and [?]
where necessary and sufficient conditions for criticality were derived for Killing
vector fields. A geometric interpretation of these conditions is yet to be found.
In this paper, we will exhibit some contact, non-Killing critical vector fields
arising from the flat contact metric geometry setting. However, our critical
non-Killing vector fields are not absolute minimizers for the volume.

Let $f(V) = \sqrt{det\,\mathcal{L}_V}$ where $det\,\mathcal{L}_V$ stands for the determinent of $\mathcal{L}_V$
and let $\Omega$ denotes the riemannian volume element on $M$. Then the volume
functional on the space of unit vector fields is given by the expression

$$\mathcal{F}(V) = \int_M f(V)\Omega. \tag{1}$$

Identifying the tangent space to $\mathcal{X}^1(M)$ at $V$ with the space of vector fields
perpendicular to $V$, one can compute the first variation of $\mathcal{F}$ as follows. For
an arbitrary vector field $A \in T_V \mathcal{X}^1(M)$, that is, for arbitrary $A$ perpendicular
to $V$, let $V(t)$ be a smooth path in $\mathcal{X}^1(M)$ such that $V(0) = V$ and $V'(0) = A$.

Denoting $\mathcal{L}_{V(t)}$ simply by $\mathcal{L}(t)$, we observe that

$$\mathcal{L}'(t) = (\nabla V'(t))^t \circ \nabla V(t) + (\nabla V(t))^t \circ \nabla V'(t)$$

and therefore, $\mathcal{L}(t)$ satisfies a linear differential equation of the type

$$X'(t) = P(t)X(t), \tag{2}$$

with

$$P(t) = [(\nabla V'(t))^t \circ \nabla V(t) + (\nabla V(t))^t \circ \nabla V'(t)] \circ \mathcal{L}^{-1}(t).$$

It follows that

$$\frac{d}{dt} det\mathcal{L}(t) = tr(P(t))det\mathcal{L}(t). \tag{3}$$

Now we can see that

$$(f \circ V)'(t) = \frac{1}{2}(det\mathcal{L}(t))^{-\frac{1}{2}}\frac{d}{dt}(det\mathcal{L}(t)). \tag{4}$$

Applying identity (??) and evaluating at $t = 0$, we obtain that the differential of $F$ at $V$ acting on the tangent vector $A$ can be written as (see $^?$, $^?$)

$$T_V\mathcal{F}(A) = \int_M f(V)tr(\mathcal{L}_V^{-1} \circ (\nabla V)^t \circ \nabla A)\Omega. \tag{5}$$

The notation $tr(.)$ in this paper stands for the trace of an operator.

## 3 The energy functional

The energy $\mathcal{E}(V)$ of a unit vector field $V$ on a closed $n$-dimensional riemannian manifold $M$ is defined as the energy in the sense of $^?$ of the map

$$V\colon (M, g) \to (T^1M, g_s),$$

where $g_s$ denotes also the restriction of the Sasaki metric on the tangent unit sphere bundle. The functional $\mathcal{E}(V)$ is given by the expression (see reference $^?$)

$$\mathcal{E}(V) = \frac{1}{2}\int_M \|\nabla V\|^2 \, \Omega + \frac{n}{2} \, Volume \, of \, M.$$

The first variation $T_V\mathcal{E}$ and the second variation $Hess_\mathcal{E}$ are given by the following formulas. For arbitrary vector fields $A$ and $B$ perpendicular to $V$,

$$T_V\mathcal{E}(A) = \int_M g(A, \nabla^*\nabla V) \, \Omega$$

$$Hess_\mathcal{E}(A,B) = \int_M g(\nabla^*\nabla A - \|\nabla V\|^2 A - R(A,\nabla V)\nabla V, B)\ \Omega,$$

where $-\nabla^*\nabla = tr(\nabla^2)$ is the trace of $\nabla^2$ and $R$ is the Riemann curvature tensor field.

## 4  Some contact metric geometry

A contact form on a $2n+1$-dimensional manifold $M$ is a 1-form $\alpha$ such that the identity $\alpha \wedge (d\alpha)^n \neq 0$ holds everywhere on $M$. Given such a 1-form $\alpha$, there is always a unique vector field $Z$ satisfying $\alpha(Z) = 1$ and $i_Z d\alpha = 0$. The vector field $Z$ is called the *characteristic vector filed* of the contact manifold $(M,\alpha)$ and the corresponding 1-dimensional foliation is called a *contact flow*. The $2n$-dimensional distribution

$$D(p) = \{X_p \in T_p M : \alpha(p)(X_p) = 0\}$$

is called the *contact distribution*. It carries a $(1,1)$ tensor field $J$ such that $-J^2$ is the identity on $D(p)$. The tensor field $J$ extends to all of $TM$ with the requirement that $JZ = 0$.

Also, any contact manifold $(M,\alpha)$ carries a nonunique riemannian metric $g$ adapted to $J$ and $\alpha$ in the sense that the identities

$$d\alpha(X,Y) = g(X,JY) \quad \text{and} \quad \alpha(X) = g(Z,X)$$

are satisfied for any vector fields $X$ and $Y$ on $M$. Such a metric $g$ is called a *contact metric* and the data $(\alpha, Z, g, J)$ is called a *contact metric structure*.

On a contact metric manifold $(M,\alpha,Z,g,J)$, one has the following identities involving the symmetric tensor field $h = \frac{1}{2}L_Z J$ ($L_Z$ denotes the usual Lie derivative of the tensor field $J$), the covariant derivative operator $\nabla$ and the curvature tensor $R$.

$$\nabla_X Z = -JX - JhX$$
$$\frac{1}{2}(R(Z,X)Z - JR(Z,JX)Z) = h^2 X + J^2 X. \tag{6}$$

When the tensor field $h$ is identitically zero, the contact metric structure is called *K*-contact. If the identity

$$(\nabla_X J)Y = g(X,Y)Z - \alpha(Y)X$$

holds for any tangent vectors $X$ and $Y$, then the contact metric structure is said to be sasakian.

## 5 The contact metric geometry of the tangent unit sphere bundle

The fundamental 1-form $\Theta$ is defined on $T^*M$ by $\Theta_\mu(v) = \mu(\pi_*v)$ where $v \in T_\mu T^*M$ and $\pi_*: TT^*M \to TM$ is the differential of the projection map $\pi: T^*M \to M$. The 2-form $\Omega = -d\Theta$ is a symplectic form on $T^*M$. We refer to $^?$ for the basics of symplectic geometry. Given a riemannian metric $g$ on $M$, the fundamental 1-form $\Theta$ pulls back to a 1-form $\tilde{\Theta}$ on $TM$, $\tilde{\Theta} = \flat^*\Theta$, where $\flat: TM \to T^*M$ is the usual musical isomorphism determined by the metric $g$. The 2-form $\tilde{\Omega} = -d\tilde{\Theta} = -\flat^*d\Theta$ is a symplectic form on $TM$ (see reference $^?$, pages 246-247). The map

$$(\pi_*, \kappa): TTM \to TM \oplus TM \tag{7}$$

is a vector bundle isomorphism along $\pi: TM \to M$, which determines an almost complex structure $J_{TM}$ on $TM$ such that, if $(\pi_*, \kappa)(x) = (u, v)$, then $(\pi_*, \kappa)(J_{TM}x) = (-v, u)$ $(^?, {}^?)$.

The tangent bundle of a riemannian manifold $M$ carries a distinguished vector field $U$ called the geodesic spray. $U$ is determined by $\pi_*U(p, v) = v$, and $\kappa U(p, v) = 0$ for any $(p, v) \in TM$. With respect to the Sasaki metric $g_s$, the 1-form $\tilde{\Theta}$ satisfies the identity $\tilde{\Theta}(V) = g_s(U, V)$ for any section $V$ of $TTM$. That is, $\tilde{\Theta}$ and $U$ are (Sasaki) metric duals. The symplectic form $\tilde{\Omega} = -d\tilde{\Theta}$ is compatible with the pair $(g_s, J_{TM})$ in the sense that the identity

$$\tilde{\Omega}(X, Y) = g_s(X, J_{TM}Y) \tag{8}$$

holds for any pair $(X, Y)$ of tangent vector fields on $TM$.

Letting $j: T^1M \to TM$ be the inclusion of the tangent unit sphere bundle as a hypersurface in $TM$, then the pulled back 1-form $\tilde{\alpha}_g = j^*\tilde{\Theta}$ is a contact form on $T^1M$ (see reference $^?$) whose characteristic vector field is tangent to the geodesic flow of $M$, its integral curves project to geodesics of $M$. The kernel of $\tilde{\alpha}_g$ has associated almost complex operator $J_{T^1M}$ determined by the equations

$$J_{T^1M}(U) = 0 \text{ and } J_{T^1M}(X) = J_{TM}X$$

for any $X$ tangent to $T^1M$ satisfying the identity $g_s(X, U) = 0$.

Under the identification (??), a vector tangent at $(p, v) \in TM$ is a couple $(u, \nabla_u W)$ where $u \in T_pM$ and $W$ is a vector field on $M$ such that $W(p) = v$. A vector tangent at $(p, v) \in T^1M$ is a couple $(u, \nabla_u W)$ where as above, $u$ is a tangent vector at $p$, but now $W$ is a unit vector field such that $W(p) = v$. Note that in this case, one has automatically the identity $g(v, \nabla_u W) = 0$. Now, if $(M, \alpha, Z, g, J)$ is a contact metric structure, a vector tangent to $Z(M) \subset T^1M$

at $(p, Z)$ is a couple $(u, \nabla_u Z)$ where $u$ is a tangent vector at $p \in M$. We also point out that under the identification (??), the vector field $U$ on $T^1 M$ is given by $U(p, v) = (v, 0)$. ¿From now on, it will be understood that the tangent unit sphere bundle is endowed with the restriction of the Sasaki metric $g_s$.

## 6 Criticality of K-contact unit vector fields

In the case of the three-dimensional sphere, the Hopf vector fields are absolute minima for the volume and the energy functionals (see references [?] and [?]). However, in higher dimensions, these vector fields are not minima anymore, but they are still critical. Their criticality is a special case of a more general result whose proof is contained in reference [?]. To state it, we need the following definition.

**Definition 1** *A submanifold $M$ in a contact manifold $(N, \alpha, J, Z)$ is said to be contact invariant if the characteristic vector field $Z$ is tangent to $M$ and $JX$ is tangent to $M$ whenever $X$ is.*

Clearly, a contact invariant submanifold inherits a contact metric structure from the ambient manifold. In this context, the inclusion of a contact invariant submanifold is a $J$-holomorphic map in the sense of reference [?] where such maps are shown to be harmonic.

**Theorem 1** *Let $(\alpha, Z, g, J)$ be contact metric structure tensors on a manifold $M$. Then the vector field $Z$ defines a contact invariant submanifold and a $J$-holomorphic embedding in $T^1 M$ if and only if $\alpha$ is a K-contact form.*

**Proof** The geodesic spray $U$ is clearly tangent to $Z(M)$. Let $(u, \nabla_u Z)$ be a tangent vector at $(p, Z) \in Z(M)$ such that $G((u, \nabla_u Z), (Z, 0)) = 0$, that is a tangent vector in the kernel of the contact form $\tilde{\alpha}_g$ on $T^1 M$. Then

$$J_{T^1 M}(u, \nabla_u Z) = (-\nabla_u Z, u) = (Ju + Jhu, u).$$

Therefore, $J_{T^1 M}(u, \nabla_u Z)$ will be tangent to $Z(M)$ if and only if $\nabla_{Ju+Jhu} Z = u$. But it is easily seen that $\nabla_{Ju+Jhu} Z = u - h^2 u$, hence $Z(M)$ will be contact invariant if and only if $h = 0$, that is, $\alpha$ is a K-contact form.

Also, observe that $Z_*(Z) = (Z, 0) = U(p, Z)$, which means that as a map between two contact manifolds, $Z$ exchanges the two characteristic vector fields involved. Let $v \in T_p M$ be a tangent vector such that $\alpha(v) = 0$. On one hand,

$$J_{T^1 M} Z_*(v) = J_{T^1 M}(\pi_* Z_*(v), \kappa Z_*(v)) = J_{T^1 M}(v, \nabla_v Z) = (Jv + Jhv, v).$$

On the other hand,

$$Z_*(Jv) = (Jv, \nabla_{Jv} Z) = (Jv, v - hv).$$

The two identities above show that $Z$ is a $J$-holomorphic map if and only if $h = 0$, that is, $\alpha$ is a K-contact form.  □

From Theorem ?? above, we derive the following corollary.

**Corollary 1** *Let $(M, \alpha, Z, g, J)$ be a K-contact structure. Then the characteristic vector field $Z$ is a minimal embedding and a harmonic map into $T^1M$. In particular, $Z$ is a minimal unit vector field and a harmonic section.*

## 7  Minimality of sasakian-Einstein unit vector fields

It was shown in the previous section that the characteristic vector fields of K-contact forms are critical for the energy and volume functionals, but in general they are not minimizing. In this section, we consider the following modified functional introduced by Brito [?].

$$D(V) = \int_M (\|\nabla V\|^2 + (n-1)(n-3)\|H_{V^\perp}\|^2)\Omega \tag{9}$$

where $H_{V^\perp}$ is the mean curvature vector field of the distribution orthogonal to $V$ and $n$ is the dimension of $M$. When $n = 3$, the functionals $E$ and $D$ coincide up to renormalizations.

Brito [?] has shown that Hopf vector fields are absolute minimizers for the functional $D$ and that in dimension 3, they are the unique minimizers. For general sasakian-Einstein manifolds, we can only prove the following.

**Proposition 1** *Let $M$ be a closed connected sasakian-Einstein $(2n+1)$-dimensional manifold with characteristic vector field $Z$. Then $Z$ is an absolute minimum for the functional $D$.*

If $M$ is a $(2n+1)$-dimensional sasakian-Einstein manifold, then $Ricci(V, V) = 2n$ for any unit vector field $V$. Therefore, from Brito's lower bound

$$D(V) \geq \int_M Ricci(V, V)\Omega, \tag{10}$$

we see that on a sasakian-Einstein $2n + 1$-manifold $M$,

$$D(V) \geq 2nVol(M). \tag{11}$$

**Lemma 1** *Let $Z$ be the characteristic vector field of a contact form on a $2n + 1$-dimensional contact metric manifold $M$. Then the mean curvature vector field $H_{Z^\perp}$ of the contact distribution is trivial.*

**Proof** We shall denote by $(g, \alpha, Z, J)$ the contact metric structure tensors. Choose a local orthonormal basis $Z, E_i, JE_i$, $i = 1, 2, ..., n$ consisting of eigenvectors for the symmetric tensor $h = \frac{1}{2}L_Z J$ with eigenvalues $\lambda_i$ and $-\lambda_i$ respectively. Then, using the first of identities (??),

$$g(H_{Z^\perp}, V) = \frac{1}{2n} \sum_{i=1}^{n} g(\nabla_{E_i} E_i + \nabla_{JE_i} JE_i, Z)$$

$$= \frac{1}{2n} \sum_{i=1}^{n} \lambda_i (g(E_i, JE_i) - g(JE_i, J^2 E_i)) = 0.$$

□

If $Z$ is the characteristic vector field of a sasakian form, then using the same orthonormal basis $Z, E_i, JE_i$, $i = 1, 2, ..., n$ as above, we see that

$$\|\nabla Z\|^2 = 2n.$$

This together with Lemma ?? shows that $D(Z) = 2n Vol(M)$ and hence $Z$ is an absolute minimizer by estimate (??).

## 8 Non-Killing critical unit vector fields

If the curvature tensor of a riemannian manifold $(M, g)$ is identically zero, then the second identity (6), with $X$ a unit tangent horizontal vector, leads to the identity

$$0 = g(hX, hX) - g(X, X) = g(hX, hX) - 1.$$

Therefore, the eigenvalues of $h$ are $\pm 1$ and the contact distribution $D$ decomposes into the positive and negative eigenbundles as $D = [+1] \oplus [-1]$.

Throughout this section, $\alpha$, $g$, $J$, and $Z$ will denote structure tensors of a flat contact metric, necessarily three-dimensional manifold $M$.

It is proven in reference [?] that $M$ admits a parallelization by three unit vector fields $Z$, $X$, and $JX$ where $Z$ and $X$ are characteristic vector fields of contact forms sharing the same flat contact metric $g$. The three vector fields $Z$, $X$ and $JX$ are respectively eigenvector fields corresponding to the eigenvalues 0, -1 and 1 of the tensor field $h$. Concerning these particular vector fields $Z$, $X$ and $JX$, the following identities were established in reference [?].

$$\nabla_Z X = 0 = \nabla_X Z, \quad \nabla_{JX} JX = 0$$
$$\nabla_{JX} Z = 2X, \quad \nabla_{JX} X = -2Z. \tag{12}$$

Moreover, the vector field $JX$ has been shown to be parallel. Still in the flat contact metric geometry setting, using identities (??) and notations of

Section 1, we see that:

$$\mathcal{L}_Z Z = Z, \quad \mathcal{L}_Z X = X, \quad \mathcal{L}_Z JX = 5JX. \tag{13}$$

From identities (??), we deduce that

$$\mathcal{L}_Z^{-1} Z = Z, \quad \mathcal{L}_Z^{-1} X = X, \quad \mathcal{L}_Z^{-1} JX = \frac{1}{5}JX. \tag{14}$$

Also, a short calculation shows that $f(Z) = \sqrt{5} = f(X)$ and $f(JX) = 1$. Hence

$$\mathcal{F}(Z) = \sqrt{5} Volume \ of \ M = \mathcal{F}(X), \quad and \quad \mathcal{F}(JX) = Volume \ of \ M. \tag{15}$$

Next, we compute the expression of $T_Z \mathcal{F}(A)$ for any horizontal vector field $A$. Using the symmetry of $\mathcal{L}_Z^{-1}$ and identities (??) and (??), we see that:

$$g(\mathcal{L}_Z^{-1} \circ (\nabla Z)^t (\nabla_Z A), Z) = g((\nabla Z)^t (\nabla_Z A), Z) = g(\nabla_Z A, \nabla_Z Z) = 0$$
$$g(\mathcal{L}_Z^{-1} \circ (\nabla Z)^t (\nabla_X A), X) = g(\nabla_X A, \nabla_X Z) = 0. \tag{16}$$

In order to complete our computation of $T_Z \mathcal{F}(A)$, we need the fact that any smooth horizontal vector field $A$ decomposes as $A = aX + bJX$ for some smooth functions $a$ and $b$ on $M$. Again, using identities (??), (??) and symmetry of $\mathcal{L}_Z^{-1}$, we find that

$$\begin{aligned} g(\mathcal{L}_Z^{-1} \circ (\nabla Z)^t (\nabla_{JX} A), JX) &= \frac{1}{5} g(\nabla_{JX} A, \nabla_{JX} Z) \\ &= \frac{2}{5} g(\nabla_{JX} A, X) \\ &= \frac{2}{5} g(da(JX)X + a\nabla_{JX} X + db(JX)JX, X) \\ &= \frac{2}{5} da(JX). \end{aligned} \tag{17}$$

¿From identities (??) and (??), we deduce that

$$tr(\mathcal{L}_Z^{-1} \circ (\nabla Z)^t \circ \nabla A) = \frac{2}{5} da(JX). \tag{18}$$

Therefore, using the fact that $JX$ is Killing, hence divergence free, we obtain

$$\begin{aligned} T_Z \mathcal{F}(A) = \frac{2}{\sqrt{5}} \int_M da(JX)\Omega &= \frac{2}{\sqrt{5}} \int_M L_{JX}(a\,\Omega) \\ &= \frac{2}{\sqrt{5}} \int_M d(i_{JX}(a\,\Omega)) = 0. \end{aligned} \tag{19}$$

(The symbol $L_{JX}$ above stands for the Lie derivative, not to be confused with $\mathcal{L}_{JX}$.)

**Remark 1** *Since Z and X play interchangeable roles in the flat contact metric setting of this section, one also has that $T_X \mathcal{F}(A) = 0$ for any A perpendicular to X.*

**Theorem 2** *Every closed, flat contact metric manifold has a parallelization by three critical unit vector fields. One of them is parallel minimizing, the other two are contact, non-Killing and do not minimize the volume functional.*

**Proof** Since $JX$ is parallel, it is obviously a minimizer for the volume functional. The last of identities (??) and the above remark show that $Z$ and $X$ are critical unit vector fields. Using identities (??), a direct computation of $L_Z g$ and $L_X g$ shows that none of $Z$ and $X$ is Killing. Finally, identities (??) show that, even though they are critical, neither $Z$, nor $X$ is minimizing for the volume functional $\mathcal{F}$.  □

Using identities (??), we easily see that

$$\nabla^* \nabla Z = 4Z.$$

Therefore, for any vector field $A$ perpendicular to $Z$, one obtains that

$$T_Z \mathcal{E}(A) = \int_M g(A, 4Z)\, \Omega = 0. \tag{20}$$

Identity (??) above shows that $Z$ is a critical vector field for the energy functional. We also have the following fact.

$$Hess_\mathcal{E}(JX, JX) = -\int_M 4\, \Omega = -4\, Volume\ of\ M\ < 0. \tag{21}$$

Identities (??) and (??) imply the following result.

**Theorem 3** *The characteristic vector field of a flat contact metric structure on a closed manifold is critical unstable for the energy functional.*

## References

1. Blair, D., *Contact manifolds in riemannian geometry*, Springer Lectures Notes in Math. **509**, Springer-Verlag, Berlin and New York, 1976.
2. Brito, F.G.B., *Total bending of flows with mean curvature correction*, Diff. Geom. Appl. **12** (2000), 157-163.
3. Dombrowski, P., *On the geometry of the tangent bundles*, J. Reine Angew. Math. **210** (1962), 73–88.
4. Eells, J., Lemaire, L., *Selected topics in harmonic maps* in CBMS **50** (1983), AMS, Providence RI.

5. Gil-Medrano, O., *On the volume functional in the manifold of unit vector fields*, Publ. Depto. Geometria y Topologia, Univ. Santiago de Compostela **89** (1998), 155–163.

6. Gil-Medrano, O., *Relationship between volume and energy of vector fields*, Diff. Geom. Appl. **15** (2001), 137-152.

7. Gil-Medrano, O., Llinares-Fuster, E., *Minimal unit vector fields*, Preprint.

8. Gluck, H., Ziller, W., *On the volume of a unit vector field on the three sphere*, Comment. Math. Helv. **30** (1986), 177–192.

9. Ianus, S., Pastore, A.M., *Harmonic maps on contact metric manifolds*, Ann. Math. Blaise Pascal **2**, No 2 (1995), 43–53.

10. Ishihara, S., Yano, K., *Tangent and cotangent bundles. Differential geometry*, Marcel Dekker Inc., New York, 1973.

11. Johnson, D.L., *Volume of flows*, Proc. Amer. Math. Soc. **104** (1988), 923-932.

12. Libermman, P., Marle, C.M., *Symplectic geometry and analytical mechanics*, Reidel, 1987.

13. Poor, W.A., *Differential geometric structures*, McGraw-Hill Book Company, New York, 1981.

14. Rukimbira, P., *A characterization of flat contact metric geometry*, Hous. Jour. Math. **24** (1998), 409–414.

15. Rukimbira, P., *Some remarks on R-contact flows*, Ann. Glob. Anal. Geom. **11** (1993), 165–171.

16. Rukimbira, P., *Criticality of K-contact vector fields*, J. Geom. Phys. **40** (2002), 209–214.

17. Tondeur, P., *Structure presque kaehlerienne naturelle sur le fibré des vecteurs covariants d'une variété riemanienne*, C.R. Acd. Sci. Paris **254** (1962), 407–408.

18. Wiegmink, G., *Total bending of vector fields on the sphere* $S^3$, Diff. Geom. Appl. **6** (1996), 219-236.

19. Wood, C.M., *On the energy of unit vector fields*, Geom. Dedicata **64** (1997), 319-330.

# ORBIFOLD HOMEOMORPHISM AND DIFFEOMORPHISM GROUPS

## JOSEPH E. BORZELLINO

*Department of Mathematics, California State Polytechnic University,*
*San Luis Obispo, CA 93407*
*E-Mail: jborzell@calpoly.edu*

## VICTOR BRUNSDEN

*Department of Mathematics and Statistics, Penn State Altoona,*
*3000 Ivyside Park, Altoona, PA 16601,*
*E-Mail: vwb2@psu.edu*

In this paper we outline results on orbifold diffeomorphism groups that were presented at the International Conference on Infinite Dimensional Lie Groups in Geometry and Representation Theory at Howard University, Washington DC on August 17-21, 2000. Specifically, we define the notion of reduced and unreduced orbifold diffeomorphism groups. For the reduced orbifold diffeomorphism group we state and sketch the proof of the following recognition result: Let $\mathcal{O}_1$ and $\mathcal{O}_2$ be two compact, locally smooth orbifolds. Fix $r \geq 0$. Suppose that $\Phi : \text{Diff}_{\text{red}}^r(\mathcal{O}_1) \to \text{Diff}_{\text{red}}^r(\mathcal{O}_2)$ is a group isomorphism. Then $\Phi$ is induced by a (topological) homeomorphism $h : X_{\mathcal{O}_1} \to X_{\mathcal{O}_2}$. That is, $\Phi(f) = hfh^{-1}$ for all $f \in \text{Diff}_{\text{red}}^r(\mathcal{O}_1)$. Furthermore, if $r > 0$, $h$ is a $C^r$ manifold diffeomorphism when restricted to the complement of the singular set of each stratum. We then show that if we replace the reduced orbifold diffeomorphism group by the unreduced orbifold diffeomorphism group in the above theorem, we can strengthen the homeomorphism $h$ to an orbifold homeomorphism (orbifold structure preserving). Lastly, we state a structure theorem for the orbifold diffeomorphism group, showing that it is a Banach manifold for $1 \leq r < \infty$ and a Fréchet manifold if $r = \infty$. As a corollary of this we obtain extensions of theorems of the second author to the setting of a smooth, compact orbifold.

## 1 Introduction

Orbifolds are a useful and interesting generalization of the notion of manifolds. They were first studied by Satake[24,25] where they were referred to as $V$-manifolds and later by Thurston.[27] In addition to being objects of study in their own right, they have come up as a new and unfamiliar domain one must pass through in order to study problems about manifolds. Some examples are the recent solution to the Arnold Conjecture by Fukaya and Ono,[20] the study of convergence of Riemannian manifolds by Anderson and Cheeger,[1] and their occurrence in problems in low dimensional topology, as in Scott.[26] One might also consider to what extent theorems about manifolds carry over to orbifolds. For example, many important theorems of Riemannian geometry have nice generalizations in the orbifold category. See Borzellino[7,8,9] and Borzellino and Zhu.[14] Of interest to us here are the appropriate orbifold versions of the recognition theorems of Filipkiewicz,[19] Rubin,[22] Rybicki[23] for manifolds given their homeomorphism (resp. diffeomorphism) groups and the refinements of these by Banyaga to recognition theorems for symplectic,[2,6] contact[3,4] and smooth unimodular[4,5] structures. In all of these recognition theorems, one of the primary ingredients needed is that the group of structure preserving transformations act transitively in some sense. In trying to extend similar results to the category of orbifolds, we immediately find that this primary ingredient is missing. Orbifolds are not homogeneous objects in the sense that manifolds without boundary are. Instead, one of the distinguishing features is that they may have a nowhere dense singular set that must be preserved under any structure preserving transformation. Even though the underlying topological space of an orbifold with non-empty singular set may be a manifold, the transformations that preserve the orbifold structure will never act transitively.

Another difficulty is that many of the fundamental notions in the manifold category do not have a unique *correct* generalization to the "category"of orbifolds. In other words, there is more than one natural category of orbifolds to work with. Despite this, however, many manifold recognition results when formulated with proper care give the appropriate analogues in the orbifold category.

Our results may be summarized by the following theorems. The first is a partial recognition theorem for orbifold structures. While not an exact analogue for the recognition theorems mentioned above, it does give an indication of the kinds of results that can be proven and the phenomena that typically arise in the orbifold category.

**Theorem A.** (Borzellino and Brunsden[10]) *Let $\mathcal{O}_1$ and $\mathcal{O}_2$ be two compact, locally smooth orbifolds. Fix $r \geq 0$. Suppose that $\Phi : \mathrm{Diff}^r_{\mathrm{red}}(\mathcal{O}_1) \to \mathrm{Diff}^r_{\mathrm{red}}(\mathcal{O}_2)$ is a group isomorphism. Then $\Phi$ is induced by a homeomorphism $h : X_{\mathcal{O}_1} \to X_{\mathcal{O}_2}$. That is, $\Phi(f) = hfh^{-1}$ for all $f \in \mathrm{Diff}^r_{\mathrm{red}}(\mathcal{O}_1)$. Furthermore, if $r > 0$, $h$ is a $C^r$ manifold diffeomorphism when restricted to the complement of the singular set of each stratum.*

Here, $\mathrm{Diff}^r_{\mathrm{red}}(\mathcal{O})$ denotes the *reduced $C^r$ orbifold diffeomorphism group* and $X_{\mathcal{O}}$ the underlying topological space of an orbifold $\mathcal{O}$. Note that in our first paper,[10] all diffeomorphisms were *reduced orbifold diffeomorphisms*. There is also a notion of *unreduced $C^r$ orbifold diffeomorphism group*, and the corresponding result using these groups is given as Theorem B below.

The restriction to compact orbifolds cannot be removed if one insists on using the reduced orbifold maps as the following example shows.

**Example 1.** Let $\mathcal{O}_1 = (0, 1)$ and $\mathcal{O}_2 = [0, 1]$, the open and closed unit intervals. These orbifolds have the same homeomorphism group, but are clearly not homeomorphic spaces.

Also, in general, the homeomorphism $h$ in Theorem A is not necessarily an orbifold homeomorphism and Theorem A is in some sense the best one can hope for if one insists on using reduced orbifold diffeomorphism groups. To see this, consider the following

**Example 2.** Let $\mathcal{O}_i, (i = 1, 2)$ be two so-called $\mathbb{Z}_{p_i}$-teardrops with $p_1 \neq p_2$. It is clear that the homeomorphism groups of $\mathcal{O}_i$ are each isomorphic to the subgroup of the homeomorphism group of the 2-sphere $S^2$ which fix the north pole. To see this, just observe that any homeomorphism of $S^2$ that fixes the north pole can be locally lifted to a $p_i$-fold covering of a neighborhood of the north pole. Note, however, that the orbifolds themselves are not *orbifold* homeomorphic, even though their underlying spaces $X_{\mathcal{O}_i} = S^2$, are topologically homeomorphic.

One might also notice that the work of Banyaga,[4] Filipkiewicz,[19] Rubin[22] and Rybicki[23] collectively show that any automorphism of the structure preserving group of transformations in the topological (with or without boundary), differentiable, PL, Lipschitz, symplectic and contact categories must be an inner automorphism. For the orbifold category, this is not the case, as the following example shows.

**Example 3.** (*Borzellino and Brunsden*[11]) For each $n > 1$ there exists a compact connected orbifold $\mathcal{O}$ of dimension $n$, such that the group of automorphisms $\mathrm{Aut}(\mathrm{Diff}^r_{\mathrm{red}}(\mathcal{O})) \neq \mathrm{Inn}(\mathrm{Diff}^r_{\mathrm{red}}(\mathcal{O}))$, the group of inner automorphisms. To see this, parameterize $S^2$ with spherical coordinates $(\theta, \phi)$, $0 \leq \theta < 2\pi$, $-\pi/2 \leq \phi \leq \pi/2$. Let $a = (\theta, -\pi/2)$ be the north pole and $b = (\theta, \pi/2)$ be the south pole. Give $S^2$ the structure of a $(p, q)$-football orbifold $\mathcal{F}$ with $p \neq q$ so the singular set $= \{a\} \cup \{b\}$. It is not hard to see that $\mathrm{Diff}^r_{\mathrm{red}}(\mathcal{F})$ is isomorphic to the group of $C^r$ diffeomorphisms of $S^2$ which fix $a$ and $b$ pointwise. Consider the group automorphism $\Phi : \mathrm{Diff}^r_{\mathrm{red}}(\mathcal{F}) \to \mathrm{Diff}^r_{\mathrm{red}}(\mathcal{F})$ defined by $(\Phi(f)) = g \circ f \circ g^{-1}$ where $g(\theta, \phi) = (\theta, -\phi)$. Then $\Phi \notin \mathrm{Inn}(\mathrm{Diff}^r_{\mathrm{red}}(\mathcal{F}))$. To see this, suppose $\Phi \in \mathrm{Inn}(\mathrm{Diff}^r_{\mathrm{red}}(\mathcal{F}))$, so that there exists $h \in \mathrm{Diff}^r_{\mathrm{red}}(\mathcal{F})$ with $\Phi(f) = \Psi(f) = h \circ f \circ h^{-1}$ for all $f \in \mathrm{Diff}^r_{\mathrm{red}}(\mathcal{F})$. Choose a neighborhood $U_a$ of $a$ with $h(U_a) \cap g(U_a) = \emptyset$, and let $f_0 \in \mathrm{Diff}^r_{\mathrm{red}}(\mathcal{F})$ with $\mathrm{supp}(f_0) \subset U_a$. Then $\Phi(f_0) = g \circ f_0 \circ g^{-1}$ has support in $g(U_a)$, a neighborhood of $b$. However, $\Psi(f_0) = h \circ f_0 \circ h^{-1}$ has support in $h(U_a)$, a neighborhood of $a$. Thus $\Phi \neq \Psi$, and we conclude that $\Phi$ is not an inner automorphism. Higher dimensional examples can be constructed by considering products with spheres $\mathcal{F} \times S^n$.

*Remark 4.* This behavior cannot occur for one-dimensional orbifolds since the only non-trivial compact 1–orbifolds are closed rays and closed intervals. The results in Borzellino and Brunsden[10] are enough to exclude such examples since they can have only $\mathbb{Z}_2$ singularities.

One can, however, show that Theorem A admits a generalization which does give the analogue of the reconstruction result that holds for manifolds. To show this, we must work with a different notion of orbifold diffeomorphism, namely the *unreduced* orbifold diffeomorphisms.

**Theorem B.** *Let $\mathcal{O}_1$ and $\mathcal{O}_2$ be two compact, locally smooth orbifolds. Fix $r \geq 0$. Suppose that $\Phi : \mathrm{Diff}^r_{\mathrm{Orb}}(\mathcal{O}_1) \to \mathrm{Diff}^r_{\mathrm{Orb}}(\mathcal{O}_2)$ is a group isomorphism. Then $\Phi$ is induced by a $C^r$ orbifold diffeomorphism $h : X_{\mathcal{O}_1} \to X_{\mathcal{O}_2}$. That is, $\Phi(f) = hfh^{-1}$ for all $f \in \mathrm{Diff}^r_{\mathrm{Orb}}(\mathcal{O}_1)$.*

Here, $\mathrm{Diff}^r_{\mathrm{Orb}}(\mathcal{O})$ denotes the *unreduced* $C^r$ orbifold diffeomorphism group.

The above results show that the algebraic structure of the homeomorphism (resp. diffeomorphism) groups determines the orbifold (but only the topological structure of the orbifold in the case of the reduced diffeomorphism group).

A related problem to consider is determining the topological structure of the diffeomorphism group of an orbifold. In the case of a compact manifold, it is well known that the group of $C^r$ diffeomorphisms is a manifold for $0 < r \leq \infty$ where the model space is the space of $C^r$ tangent vector fields on $M$. See, for example Banyaga.[4] This is a Banach space for $0 < r < \infty$ and a Fréchet space for $r = \infty$. One might naively think that the orbifold diffeomorphism group is itself an infinite dimensional orbifold, but one only need remember that the orbifold diffeomorphism group is a group and hence is homogeneous. Thus, it cannot be a non-trivial orbifold. In fact, in the case of a smooth compact orbifold, the structure of the orbifold diffeomorphism group holds no surprises.

**Theorem C.** *Let $\mathcal{O}$ be a smooth compact orbifold without boundary and let $\mathrm{Diff}^r_{\mathrm{Orb}}(\mathcal{O})$ be the group of unreduced $C^r$ orbifold diffeomorphisms equipped with the topology of uniform convergence of all derivatives of orders $\leq r$. Then $\mathrm{Diff}^r_{\mathrm{Orb}}(\mathcal{O})$ is a manifold modeled on the topological vector space $D^r_{\mathrm{Orb}}(\mathcal{O})$ of $C^r$ orbifold sections of the tangent orbibundle equipped with the topology of uniform convergence of derivatives of order $\leq r$. This vector space is a Banach space if $r < \infty$ and is a Fréchet space if $r = \infty$.*

The rest of the paper will be devoted primarily to giving the background information and terminology used the statements of the above theorems. We will also give an indication of the proofs of Theorem A and Theorem B. The full proof of Theorem A can be found in our first paper[10] and the proofs of Theorem B and Theorem C will appear elsewhere.[12,13] Since it is a fundamental tool in proving both Theorems A and B, we recall the following theorem of Rubin.[22] A subset $S$ of a topological space $X$ is called *somewhere dense* if the interior of its closure is nonempty. That is, $\text{int}(\text{cl}(S)) \neq \emptyset$.

**Theorem (Rubin).** *Let $X_i, (i = 1, 2)$ be locally compact Hausdorff spaces and $G_i$ subgroups of the group of homeomorphisms of $X_i$ such that for every open set $T \subset X_i$ and $x \in T$ the set $\{g(x) \mid g \in G_i \text{ and } g\mid_{(X_i - T)} = \text{Id}\}$ is somewhere dense. Then if $\Phi : G_1 \to G_2$ is a group isomorphism, then there is a homeomorphism $h$ between $X_1$ and $X_2$ such that for every $g \in G_1$, $\Phi(g) = hgh^{-1}$.*

As a Corollary of the above results, we obtain generalizations of results of the second author (see Brunsden[16,17]) to actions of finitely generated groups on smooth compact orbifolds.

**Corollary 5.** *Let $\mathcal{O}$ be a smooth compact orbifold, $\Gamma$ a finitely generated group and $\phi \in \text{Hom}(\Gamma, \text{Diff}^r_{\text{Orb}}(\mathcal{O}))$ an action of $\Gamma$ on $\mathcal{O}$ by $C^r$ orbifold diffeomorphisms $(r > 1)$. If $H^1(\Gamma, D^{r-1}_{\text{Orb}}(\mathcal{O})) = 0$, then there is a neighborhood $U$ of $\phi$ in $\text{Hom}(\Gamma, \text{Diff}^r_{\text{Orb}}(\mathcal{O}))$ (equipped with the compact-open topology) so that for each $\psi \in U$, there is an $h \in \text{Diff}^{r-1}_{\text{Orb}}(\mathcal{O})$ so that*

$$\psi(\gamma) \circ h = h \circ \phi(\gamma)$$

*for all $\gamma \in \Gamma$. If $r = 1$, then we require that in addition $H^0(\Gamma, D^0_{\text{Orb}}(\mathcal{O})) = 0$. Here, $D^0_{\text{Orb}}(\mathcal{O})$ is as in the proof of Theorem C and is a $\Gamma$ module via the induced orbibundle map $\gamma_* : D^r_{\text{Orb}}(\mathcal{O}) \to D^r_{\text{Orb}}(\mathcal{O})$.*

## 2 Orbifolds

Our definition is modeled on the definition found in Thurston.[27]

**Definition 6.** A (topological) *orbifold* $\mathcal{O}$, consists of a paracompact, Hausdorff topological space $X_{\mathcal{O}}$ called the *underlying space*, with the following local structure. For each $x \in X_{\mathcal{O}}$ and neighborhood $U$ of $x$, there is a neighborhood $U_x \subset U$, an open set $\tilde{U}_x \cong \mathbb{R}^n$, a finite group $\Gamma_x$ acting continuously and effectively on $\tilde{U}_x$ which fixes $0 \in \tilde{U}_x$, and a homeomorphism $\phi_x : \tilde{U}_x/\Gamma_x \to U_x$ with $\phi_x(0) = x$. These actions are subject to the condition that for a neighborhood $U_z \subset U_x$ with corresponding $\tilde{U}_z \cong \mathbb{R}^n$, group $\Gamma_z$ and homeomorphism $\phi_z : \tilde{U}_z/\Gamma_z \to U_z$, there is an embedding $\tilde{\psi} : \tilde{U}_z \to \tilde{U}_x$ and an injective homomorphism $f : \Gamma_z \to \Gamma_x$ so that $\tilde{\psi}$ is equivariant with respect to $f$ (that is, for $\gamma \in \Gamma_z, \tilde{\psi}(\gamma y) = f(\gamma)\tilde{\psi}(y)$ for all $y \in \tilde{U}_z$), such that the following diagram commutes:

$$
\begin{array}{ccc}
\tilde{U}_z & \xrightarrow{\ \tilde{\psi}\ } & \tilde{U}_x \\
\downarrow & & \downarrow \\
\tilde{U}_z/\Gamma_z & \xrightarrow{\psi=\tilde{\psi}/\Gamma_z} & U_x/f(\Gamma_z) \\
& & \downarrow \\
\downarrow{\phi_x} & & \tilde{U}_x/\Gamma_x \\
& & \downarrow{\phi_x} \\
U_z & \xrightarrow{\hspace{2cm}} & U_z
\end{array}
$$

The covering $\{U_x\}$ of $X_{\mathcal{O}}$ is not an intrinsic part of the orbifold structure. We regard two coverings to give the same orbifold structure if they can be combined to give a larger covering still satisfying the definitions.

Let $0 \le r \le \infty$. An orbifold $\mathcal{O}$ is a $C^r$ *orbifold* if each $\Gamma_x$ acts $C^r$-smoothly and the embedding $\tilde{\psi}$ is $C^r$.

**Definition 7.** We say that an orbifold $\mathcal{O}$ is *locally smooth* if the action of $\Gamma_x$ on $\tilde{U}_x \cong \mathbb{R}^n$ is an *orthogonal* action for all $x \in \mathcal{O}$. That is, for each $x \in \mathcal{O}$, there exists a representation $L : \Gamma_x \to \mathbb{O}(n)$ such that if $\gamma \cdot y$ denotes the $\Gamma_x$ action on $\tilde{U}_x$, then we have $\gamma \cdot y = L(\gamma)y$ for all $y \in \tilde{U}_x$.

**Definition 8.** An *orbifold chart* about $x$ in a locally smooth orbifold $\mathcal{O}$ is a 4-tuple $(\tilde{U}_x, \Gamma_x, \rho_x, \phi_x)$ where $\tilde{U}_x = \mathbb{R}^n$, $\Gamma_x$ is a finite group, $\rho_x$ is a representation of $\Gamma_x : \rho_x \in \mathrm{Hom}(\Gamma_x, \mathbb{O}(n))$, with $\mathbb{O}(n)$ the orthogonal group, and $\phi_x$ is a homeomorphism: $\phi_x : \tilde{U}_x/\rho_x(\Gamma_x) \to U_x$, where $U_x \subset X_{\mathcal{O}}$ is a (sufficiently small) open relatively compact neighborhood of $x$, and $\phi_x(0) = x$.

For convenience we will often refer to the neighborhood $U_x$ as an orbifold chart, and ignore the representation $\rho_x$ and write $U_x = \tilde{U}_x/\Gamma_x$. If necessary we will denote by $\pi_x : \tilde{U} \to \tilde{U}/\rho_x(\Gamma_x)$, the quotient map defined by the action of $\rho_x(\Gamma_x)$ on $\tilde{U}$.

In the remainder, all orbifolds will be assumed to be locally smooth.

**Definition 9.** Let $\mathcal{O}$ be a connected $n$-dimensional locally smooth orbifold. Given a point $x \in \mathcal{O}$, there is a neighborhood $U_x$ of $x$ which is homeomorphic to a quotient $\tilde{U}_x/\Gamma_x$ where $\tilde{U}_x$ is homeomorphic to $\mathbb{R}^n$ and $\Gamma_x$ is a finite group acting orthogonally on $\mathbb{R}^n$. The definition of orbifold implies that the germ of this action in a neighborhood of the origin of $\mathbb{R}^n$ is unique. We define the *isotropy group of* $x$ to be the group $\Gamma_x$. The *singular set* of $\mathcal{O}$ is the set of points $x \in \mathcal{O}$ with $\Gamma_x \ne \{1\}$. Denote the singular set of $\mathcal{O}$ by $\Sigma_1$. Then $\Sigma_1$ is also a (possibly disjoint) union $\bigcup_{l_1} \Sigma^{(l_1)}$ of connected locally smooth orbifolds of strictly lower dimension (though different components may have different dimensions). See the section of examples. Each of the orbifolds $\Sigma_1^{(l_1)}$ has a singular set $\bigcup_{l_2} \Sigma_1^{(l_1)(l_2)}$. Define the singular set of $\Sigma_1$ to be $\Sigma_2 = \bigcup_{(l_1)(l_2)} \Sigma_1^{(l_1)(l_2)}$. Proceeding inductively, we get a stratification of $\mathcal{O}$:

$$\mathcal{O} = \Sigma_0 \supset \Sigma_1 \supset \Sigma_2 \supset \cdots \Sigma_{k-1} \supset \Sigma_k = \emptyset \text{ for some } k \leq n+1$$

By a result of M.H.A Newman,[18] we note that the singular set of a topological orbifold is a closed nowhere dense set. See also Thurston.[27]

Products of (locally smooth) orbifolds inherit a natural (locally smooth) orbifold structure:

**Definition 10.** Let $\mathcal{O}_i$ for $i = 1, 2$ be orbifolds. The *orbifold product* $\mathcal{O}_1 \times \mathcal{O}_2$ is the orbifold having the following structure:

1. $X_{\mathcal{O}_1 \times \mathcal{O}_2} = X_{\mathcal{O}_1} \times X_{\mathcal{O}_2}$.

2. For each $(x,y) \in X_{\mathcal{O}_1 \times \mathcal{O}_2}$ and pair of orbifold charts $U_x \ni x$ and $V_y \ni y$ $U_x \times V_y$ is an orbifold chart around $(x,y)$. Explicitly,

$$(\tilde{U}_x \times \tilde{V}_y, \Gamma_x \times \Gamma_y, \rho_x \times \rho_y, \phi_x \times \phi_y)$$

is an orbifold chart around $(x,y)$.

Note that the isotropy group $\Gamma_{(x,y)} = \Gamma_x \times \Gamma_y$.

We close this section with some elementary results on orbifolds. The (rather trivial) proofs can be found in our paper,[10] so we omit them here.

**Proposition 11.** *If $\mathcal{O}$ is locally smooth then in each local orbifold chart $\tilde{U}_x$ the fixed point set $S_x = \{y \in \tilde{U}_x \mid \Gamma_x \cdot y = y\}$ is a topological sub-manifold of $\tilde{U}_x$.*

**Proposition 12.** *If $\mathcal{O}$ is a smooth $C^r$ orbifold with $r > 0$, then it is locally smooth.*

## 3 Examples of Orbifolds

**Example 13.** Let $\mathcal{O} = (S^n, \text{can})/G$, $n > 1$, be the $n$–dimensional hemisphere of constant curvature 1 (topologically $\mathcal{O}$ is just the closed $n$–disk $D^n$). $G = \mathbb{Z}_2 \subset \mathbb{O}(n+1)$ is the group generated by reflection through an equatorial $(n-1)$–sphere. In this case $\Sigma_1$ is the equatorial $(n-1)$–sphere.

**Example 14.** Let $\mathcal{O}$ be a $\mathbb{Z}_p$–football. $\mathcal{O} = (S^2, \text{can})/G$, where $G \subset \mathbb{O}(3)$ is rotation around the $z$–axis in $\mathbb{R}^3$, through an angle of $2\pi/p$. Here $\Sigma_1 = \{\text{north pole}\} \cup \{\text{south pole}\}$.

**Example 15.** Let $\mathcal{O}$ be a $\mathbb{Z}_p$–football/$G$, where $G$ is reflection in the equator of the football that does not contain the singular points. Topologically, $\mathcal{O}$ is

$D^2$. Note that the singular set $\Sigma_1 = \{\text{equator}\} \cup \{\text{point}\}$, thus it is possible for different components of the singular set to have different dimensions.

**Example 16.** Let $\mathcal{O} = \mathbb{R}^2/G$, where $G$ is the crystallographic group generated by reflecting an equilateral triangle or square in each of its sides to produce a tiling of $\mathbb{R}^2$. Then $\mathcal{O}$ is just the closed triangle or square, with singular set the boundary of the tiling region. The stratification of $\mathcal{O}$ is as follows:

$$\mathcal{O} = \Sigma_0 \supset \Sigma_1 = \{\text{the boundary of the triangle or square}\} \supset$$

$$\Sigma_2 = \{\text{the vertices}\} \supset \Sigma_3 = \emptyset$$

Here, $\Sigma_1$ is union of the closed line segments making up the boundary of the triangle or square and each of these line segments is a 1–dimensional orbifold with 2 singular points. One should observe that $\Sigma_1$ is not a 1–dimensional orbifold but a *union* of 1–dimensional orbifolds. The lowest dimensional stratum has dimension 0. Note that the manifold $\Sigma_1 - \Sigma_2$ is a union of open line segments. If one only quotients out by the index 2 subgroup $G_0$ of orientation preserving elements of $G$ then $\mathcal{O}$ becomes topologically a 2–sphere. The complement of the singular set is topologically $\mathbb{R}^2 - \{2 \text{ points or } 3 \text{ points}\}$

**Example 17.** Let $\mathcal{O}$ be a $\mathbb{Z}_p$-teardrop. The underlying space of this orbifold is $S^2$ with a single conical singularity of order $p$ at the north pole.

**Example 18.** Consider the group $G = \mathbb{Z}_2 \times \mathbb{Z}_2$ generated by rotations of $\pi$ radians about the three coordinate axes of $\mathbb{R}^3$. If we consider the quotient of the 2–sphere $S^2/G$, we get a 2–dimensional orbifold $\mathcal{O}$ whose underlying space is topologically the 2–sphere with 3 singular points. The sin–suspension $\Sigma_{\text{sin}}\mathcal{O} = S^3/\Sigma G$ is an orientable 3–dimensional orbifold. $\Sigma G$ denotes the suspension of the action on $S^2$ to $S^3$. In this case, $\Sigma_1$ is the union of the 3 line segments joining the suspension points and passing through one of the singular points of $\mathcal{O}$. $\Sigma_2$ is just the two suspension points.

**Example 19.** Let $L_p = S^3/G$ be a 3–dimensional lens space. Suspend the action of $G$ to an action $\Sigma G$ on the 4–sphere $S^4$. Let $\mathcal{O} = S^4/\Sigma G$. Then the underlying space of $\mathcal{O}$ is *not* a manifold (or manifold with boundary).

**Example 20.** An $n$-dimensional smooth manifold with corners (that is, a paracompact, Hausdorff space locally modeled on $(-\infty, \infty)^k \times [0, \infty)^{n-k}$, $0 \le k \le n$, $k$ may vary from point to point, and with smooth overlaps on the charts) is an orbifold with local model $\mathbb{R}^n/G$ where $G = (\mathbb{Z}_2)^{n-k}$ and the action of $G$ is generated by reflection through the appropriate coordinate planes $x_{i_\ell} = 0$ for $\ell = 1, \ldots n - k$. The singular set $\Sigma$ is then the boundary (those points that do not have neighborhoods homeomorphic to $\mathbb{R}^n$), and the stratification given by the fixed point sets of the various subgroups of of $G$.

## 4  Orbifold Maps

We now discuss two natural definitions of maps between orbifolds. Note that in our first paper,[10] all maps discussed were *reduced orbifold maps*.

**Definition 21.** An *unreduced orbifold map* $(f, \Theta_{f,x}, \overline{\tilde{f}}_x)$ from $\mathcal{O}_1$ to $\mathcal{O}_2$ consists of the following:

1. A continuous map $f : X_{\mathcal{O}_1} \to X_{\mathcal{O}_2}$ of the underlying topological spaces.

2. For each $x$, a group homomorphism $\Theta_{f,x} : \Gamma_x \to \Gamma_{f(x)}$

3. A germ $\overline{\tilde{f}}_x$ at $0$ of a $\Theta_{f,x}$ equivariant lift $\tilde{f}_x : \tilde{U}_x \to \tilde{V}_{f(x)}$ where $(\tilde{U}_x, \Gamma_x, \rho_x, \phi_x)$ is an orbifold chart about $x$, $(\tilde{V}_{f(x)}, \Gamma_{f(x)}, \rho_{f(x)}, \phi_{f(x)})$ is an orbifold chart about $f(x)$, and such that the following diagram commutes:

$$
\begin{array}{ccc}
\tilde{U}_x & \xrightarrow{\quad \tilde{f}_x \quad} & \tilde{V}_{f(x)} \\
\downarrow & & \downarrow \\
\tilde{U}_x/\Gamma_x & \xrightarrow{\ \tilde{f}_x/\Theta_{f,x}(\Gamma_x)\ } & \tilde{V}_{f(x)}/\Theta_{f,x}(\Gamma_x) \\
\downarrow & & \downarrow \\
& & \tilde{V}_{f(x)}/\Gamma_{f(x)} \\
\downarrow & & \downarrow \\
U_x & \xrightarrow{\quad f \quad} & V_{f(x)}
\end{array}
$$

Two unreduced orbifolds maps $(f, \Theta_{f,x}, \overline{\tilde{f}}_x)$ and $(g, \Theta'_{g,x}, \overline{\tilde{g}}_x)$ are considered the same if $f = g$, $\Theta_{f,x} = \Theta'_{g,x}$, and $\overline{\tilde{f}}_x = \overline{\tilde{g}}_x$ as germs at $0$.

**Definition 22.** A *reduced orbifold map* is a continuous map $f : X_{\mathcal{O}_1} \to X_{\mathcal{O}_2}$ for which such local liftings exist. We ignore the particular choice of local lift $\tilde{f}_x$ and the choice of homomorphism $\Theta_{f,x}$.

**Definition 23.** An orbifold map $f : \mathcal{O}_1 \to \mathcal{O}_2$ (either reduced or unreduced) of smooth orbifolds is $C^r$-*smooth* if each of the local lifts $\tilde{f}_x$ may be chosen to be $C^r$.

Given two orbifolds $\mathcal{O}_i$, $i = 1, 2$, the class of $C^r$ unreduced orbifold maps from $\mathcal{O}_1$ to $\mathcal{O}_2$ will be denoted by $C^r_{\mathrm{Orb}}(\mathcal{O}_1, \mathcal{O}_2)$ and the class of reduced $C^r$ orbifold maps by $C^r_{\mathrm{red}}(\mathcal{O}_1, \mathcal{O}_2)$. For the purely topological categories of locally smooth orbifolds and unreduced (respectively reduced) continuous orbifold maps we write $C^0_{\mathrm{Orb}}(\mathcal{O}_1, \mathcal{O}_2)$ (respectively, $C^0_{\mathrm{red}}(\mathcal{O}_1, \mathcal{O}_2)$).

It is a simple matter to verify that composition of orbifold maps whether reduced or unreduced results in an orbifold map of the same type.

**Definition 24.** For any topological space $X$, let $H(X)$ denote its group of homeomorphisms. For a topological orbifold $\mathcal{O}$, the group of *unreduced orbifold homeomorphisms*, $H_{\mathrm{Orb}}(\mathcal{O})$ will be the subgroup of $H(X_{\mathcal{O}})$ so that $f, f^{-1} \in C^0_{\mathrm{Orb}}(X_{\mathcal{O}}, X_{\mathcal{O}})$. If $\mathcal{O}$ is a $C^r$ orbifold, $\mathrm{Diff}^r_{\mathrm{Orb}}(\mathcal{O})$ is the subgroup of $H_{\mathrm{Orb}}(\mathcal{O})$ with $f, f^{-1} \in C^r_{\mathrm{Orb}}(\mathcal{O})$. We will also use $\mathrm{Diff}^0_{\mathrm{Orb}}(\mathcal{O})$ for $H_{\mathrm{Orb}}(\mathcal{O})$. The corresponding notions for reduced orbifold maps will be denoted by $\mathrm{Diff}^r_{\mathrm{red}}(\mathcal{O})$.

One would expect that orbifold diffeomorphisms preserve the singular set. That is, in fact, the case and we state it here for completeness. See Borzellino and Brunsden[10] for the details of the proof.

**Lemma 25.** *Any element of* $\mathrm{Diff}^r_{\mathrm{Orb}}(\mathcal{O})$ *or* $\mathrm{Diff}^r_{\mathrm{red}}(\mathcal{O})$ *leaves* $\Sigma_i$ *invariant (as a set), where* $\Sigma_i$ *is any substratum of* $\mathcal{O}$.

The proof of Theorem A requires that apply Rubin's theorem to the complement of the singular set of an orbifold. In order to do this, we need to know that orbits of points under $\mathrm{Diff}^r_{\mathrm{red}}(\mathcal{O})$ are somewhere dense. See the previously cited paper[10] for the details.

**Lemma 26.** *The following are equivalent:*

1. $x \in \mathcal{O} - \Sigma$

2. *The orbit* $\mathrm{Diff}^r_{\mathrm{red}}(\mathcal{O}) \cdot x = \{g(x) \mid g \in \mathrm{Diff}^r_{\mathrm{red}}(\mathcal{O})\}$ *is somewhere dense.*

There is an obvious forgetful homomorphism $\tau : C^r_{\mathrm{Orb}}(\mathcal{O}_1, \mathcal{O}_2) \to C^r_{\mathrm{red}}(\mathcal{O}_1, \mathcal{O}_2)$ from the class of unreduced orbifold maps to the class of reduced orbifold maps and each reduced orbifold map comes from at least one such unreduced orbifold map.

**Definition 27.** For a $C^r$ orbifold $\mathcal{O}$, and integer $0 \le s \le r \le \infty$, let $S^s(\mathcal{O}) = \ker(\tau : \mathrm{Diff}^s_{\mathrm{Orb}}(\mathcal{O}) \to \mathrm{Diff}^s_{\mathrm{red}}(\mathcal{O})) = \{f \in \mathrm{Diff}^s_{\mathrm{Orb}}(\mathcal{O}) \mid \tau(f) = \mathrm{Id}_{\mathcal{O}}\}$. Since $S^s(\mathcal{O})$ consists of those unreduced lifts of the (always) $C^r$ smooth identity map, we see that $S^s(\mathcal{O})$ is independent of $s$, and thus we denote it more simply by $S(\mathcal{O}) = S^s(\mathcal{O}) = S^r(\mathcal{O})$, for all $0 \le s \le r \le \infty$.

The following result gives the structure of $S(\mathcal{O})$, and its proof basically follows from the definitions. A complete proof of this result will appear another article.[13]

**Proposition 28.** *Let* $x \in \mathcal{O}$, *and let* $(\tilde{U}_x, \Gamma_x, \rho_x, \phi_x)$ *be an orbifold chart around* $x$. *If* $\sigma = (\mathrm{Id}, \Theta_{\mathrm{Id},x}, \overline{\mathrm{Id}}) \in S(\mathcal{O})$, *then there is a* $\delta \in \Gamma_x$ *so that* $\overline{\mathrm{Id}}(\tilde{y}) = \delta \cdot \tilde{y}$ *for all* $\tilde{y} \in \tilde{U}_x$ *and* $\Theta_{\mathrm{Id},x}(\gamma) = \delta \cdot \gamma \cdot \delta^{-1}$. *Note that if* $x \in \mathcal{O}$ *is non-singular, then* $\delta = e$.

# 5 Orbifold Bundles, Orbibundles and Suborbifolds

Since we will ultimately want to look at sections of the tangent bundle to a smooth orbifold in order to state Theorem C, we now define the notions of unreduced and reduced orbifold bundles.

**Definition 29.** An *unreduced* (respectively, *reduced*) *orbifold bundle* is a triple $(\mathcal{E}, \mathcal{B}, p)$ where $\mathcal{E}$ and $\mathcal{B}$ are locally smooth orbifolds with $p : \mathcal{E} \to \mathcal{B}$ an unreduced (respectively, reduced) orbifold map. An orbifold bundle is *linear* if the orbifold structures on the total space $\mathcal{E}$ and base space $\mathcal{B}$ are compatible with the following local triviality conditions

1. For each $x \in X_{\mathcal{B}}$ with isotropy group $\Gamma_x$ and orbifold chart $U_x \subset X_{\mathcal{B}}$ containing $x$, so that $\tilde{U}_x \cong \mathbb{R}^n$ and $\tilde{U}_x / \Gamma_x \cong U_x$ we have $\tilde{p}^{-1}(\tilde{U}_x) \cong \tilde{U}_x \times \mathbb{R}^k$. Also there is a group $G_x$, an action $P_x \in \mathrm{Hom}(G_x, \mathbb{O}(n+k))$ and a surjective group homomorphism $\Theta_{p,(x,0)} : G_x \to \Gamma_x$ so that $\tilde{p}$ is $\Theta_{p,(x,0)}$ equivariant, i.e.:

$$\rho_x(\Theta_{p,(x,0)}(g))\tilde{p}(y) = \tilde{p}(P_x(g)y)$$

for all $g \in G_x$ and $y \in \tilde{p}^{-1}(\tilde{U}_x)$.

2. Given another $U_z \subset U_x$ with corresponding $\tilde{U}_z \cong \mathbb{R}^n$, group $\Gamma_z$, homeomorphism $\phi_z : \tilde{U}_z / \Gamma_z \to U_z$ and embedding $\tilde{\psi} : \tilde{U}_z \hookrightarrow \tilde{U}_x$ there are injective group homomorphisms $\theta_{z,x} : \Gamma_z \to \Gamma_x$, $\Theta_{z,x} : G_z \to G_x$ and embeddings $\tilde{\Psi} : \tilde{U}_z \times \mathbb{R}^k \to \tilde{U}_x \times \mathbb{R}^k$ and $\Psi : p^{-1}(U_z) \hookrightarrow p^{-1}(U_x)$ so that the following diagram commutes. (Note that all the vertical arrows are quotient maps obtained by modding out by the action of the appropriate

isotropy groups).

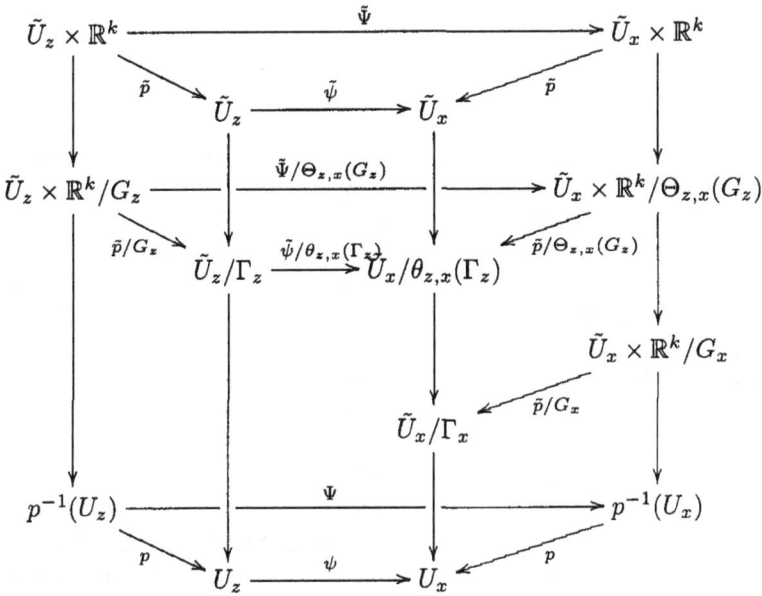

$$
\begin{array}{ccc}
\tilde{U}_z \times \mathbb{R}^k & \xrightarrow{\quad\tilde{\Psi}\quad} & \tilde{U}_x \times \mathbb{R}^k \\
\Big\downarrow \tilde{p} & \tilde{U}_z \xrightarrow{\tilde{\psi}} \tilde{U}_x & \Big\downarrow \tilde{p} \\
\tilde{U}_z \times \mathbb{R}^k/G_z & \xrightarrow{\tilde{\Psi}/\Theta_{z,x}(G_z)} & \tilde{U}_x \times \mathbb{R}^k/\Theta_{z,x}(G_z) \\
\tilde{p}/G_z & \tilde{U}_z/\Gamma_z \xrightarrow{\tilde{\psi}/\theta_{z,x}(\Gamma_z)} \tilde{U}_x/\theta_{z,x}(\Gamma_z) & \tilde{p}/\Theta_{z,x}(G_z) \\
 & & \tilde{U}_x \times \mathbb{R}^k/G_x \\
 & \tilde{U}_x/\Gamma_x & \tilde{p}/G_x \\
p^{-1}(U_z) & \xrightarrow{\quad\Psi\quad} & p^{-1}(U_x) \\
\Big\downarrow p & U_z \xrightarrow{\psi} U_x & \Big\downarrow p
\end{array}
$$

3. The mappings $\tilde{\Psi}(y,\cdot) : \tilde{p}^{-1}(y) \to \tilde{p}^{-1}(\tilde{\psi}(y))$ are invertible linear maps on $\tilde{p}^{-1}(y)$ for each $y \in \tilde{U}_z$.

In most cases the actions $P_x$ above will be in $\mathbb{O}(n) \times \mathbb{O}(k)$ rather than the more general case of $\mathbb{O}(n+k)$ that we allow above.

**Definition 30.** A *linear orbibundle* is a linear orbifold bundle with the property that for all $x \in \mathcal{O}$ one has $G_x \cong \Gamma_x$.

Suppose that one is given a topological space $X_\mathcal{E}$, a base orbifold $\mathcal{B}$, and a map $p : X_\mathcal{E} \to \mathcal{B}$ satisfying conditions (1)–(3). Then $X_\mathcal{E}$ can be given the structure as the total space of a linear orbifold bundle $p : \mathcal{E} \to \mathcal{B}$. In particular, the total space $\mathcal{E}$ is an orbifold.

**Definition 31.** Let $\mathcal{O}$ be an $n$–dimensional smooth orbifold. The *tangent orbibundle* of $\mathcal{O}$, $p : T\mathcal{O} \to \mathcal{O}$ is the linear orbibundle defined by the following construction. If $U_x$ is an orbifold chart around $x \in \mathcal{O}$, $\Gamma_x$ the isotropy group of $x$, and $\tilde{U}_x \cong \mathbb{R}^n$ so that $U_x \cong \tilde{U}_x/\Gamma_x$, then $p^{-1}(U_x) \cong (\tilde{U}_x \times \mathbb{R}^n)/\Gamma_x$ where $\Gamma_x$ acts on $\tilde{U}_x \times \mathbb{R}^n$ via

$$
\gamma \cdot (x,v) = (\gamma \cdot x, d\gamma_x(v))
$$

The definition of the tangent orbibundle allows the following

**Definition 32.** For a $C^r$ orbifold map $f : \mathcal{O}_1 \to \mathcal{O}_2$, the *tangent mapping* $Tf : T\mathcal{O}_1 \to T\mathcal{O}_2$ is the $C^{r-1}$ orbifold map defined by:

$$\widetilde{Tf}(x, v) = [(\tilde{f}(\tilde{x}), d\tilde{f}_{\tilde{x}}(\tilde{v})]$$

where $[(\tilde{x}, \tilde{v})]$ is the equivalence class of $(\tilde{x}, \tilde{v}) \in \tilde{U}_x \times \mathbb{R}^n$ and $(x, v) \cong [(\tilde{x}, \tilde{v})]$. One may similarly define the unreduced orbifold tangent mapping of an unreduced orbifold map by including the homomorphism information. $Tf$ is an orbifold map in the sense of Definition 21.

**Definition 33.** An *unreduced orbisection* of an orbifold bundle $\mathcal{E}$ over an orbifold $\mathcal{O}$ is an unreduced orbifold map $f : \mathcal{O} \to \mathcal{E}$ such that $p \circ f = \mathrm{Id}_\mathcal{O}$. In other words, it is simply a section in the category of orbibundles and unreduced orbifold maps. Likewise, we may define a reduced orbisection of a reduced orbibundle by using reduced orbifold maps.

We have the following structure result for orbisections. We defer the proof to our preprint.[13]

**Proposition 34.** *The set $C^0_{\mathrm{Orb}}(\mathcal{E})$ of unreduced orbisections of a linear orbibundle $\mathcal{E}$ is naturally a real vector space.*

**Example 35.** Given an orbifold $\mathcal{O}$, the orbifold product $\mathcal{O} \times \mathcal{O}$ is itself an orbifold. The projections $p_1$ and $p_2$ onto the first and second factors respectively give two different orbifold bundle structures to $\mathcal{O} \times \mathcal{O}$. Note that this bundle has a canonical orbifold section, the *diagonal* $= \Delta(\mathcal{O}) \subset \mathcal{O} \times \mathcal{O}$, where $\Delta : \mathcal{O} \to \mathcal{O} \times \mathcal{O}$ is defined by the diagonal map $\Delta(x) = (x, x)$ for all $x \in \mathcal{O}$ and $\Theta_{\Delta, x}(\gamma) = \gamma \times \gamma$ for all $\gamma \in \Gamma_x$.

We now consider suborbifolds. The definition of a suborbifold is somewhat more delicate than the corresponding notion for a manifold. We want a definition that is sufficiently flexible so that, in particular, the diagonal $\Delta(\mathcal{O}) \subset \mathcal{O} \times \mathcal{O}$ is a suborbifold of $\mathcal{O} \times \mathcal{O}$.

**Definition 36.** A *suborbifold* $\mathcal{P}$ of an orbifold $\mathcal{O}$ consists of the following.

1. A subspace $X_\mathcal{P} \subset X_\mathcal{O}$ equipped with the subspace topology

2. For each $x \in X_\mathcal{P}$ and neighborhood $W$ of $x$ in $X_\mathcal{P}$ there is an orbifold chart $(\tilde{U}_x, \Gamma_x, \rho_x, \phi_x)$ about $x$ in $\mathcal{O}$ with $U_x \subset W$, a subgroup $G_x \subset \Gamma_x$ of the isotropy group of $x$ in $\mathcal{O}$ and a $\rho_x(G_x)$ invariant vector subspace $\tilde{V}_x \subset \tilde{U}_x = \mathbb{R}^n$, so that

$$(\tilde{V}_x, G_x, \rho_x|_{G_x}, \psi_x)$$

is an orbifold chart for $\mathcal{P}$ and

3.

$$V_x = \psi_x(\tilde{V}_x/\rho_x(G_x))$$
$$= U_x \cap X_{\mathcal{P}}$$
$$= \phi_x(\pi_x(\tilde{V}_x))$$

is an orbifold chart for $x$ in $\mathcal{P}$ where $\pi_x : \tilde{U}_x \to \tilde{U}_x/\rho_x(\Gamma_x)$ is the quotient map.

*Remark 37.* If one only requires (1) and (2) for a suborbifold $\mathcal{P}$, then we will say that the suborbifold $\mathcal{P}$ is *immersed*.

*Remark 38.* It is tempting to define the notion of an $m$–suborbifold $\mathcal{P}$ of an $n$–orbifold $\mathcal{O}$ simply by requiring $\mathcal{P}$ to be locally modeled on $\mathbb{R}^m \subset \mathbb{R}^n$ modulo finite groups. That is, the local action on $\mathbb{R}^m$ is induced by the local action on $\mathbb{R}^n$. See Thurston[27] for a definition that takes this approach. This is equivalent to the added condition in our definition that $G_x = \Gamma_x$ at all $x$ in the underlying topological space of $\mathcal{P}$. By analogy with the definition of neat submanifolds of manifolds with boundary (see for example Hirsch[21]), we call such suborbifolds *neat*. Neat suborbifolds turn out to be too restrictive a notion for our purposes. For example the diagonal $\Delta_{\mathcal{O}} \subset \mathcal{O} \times \mathcal{O}$ is *not* a neat suborbifold of $\mathcal{O} \times \mathcal{O}$. However, using the more general definition 36, the diagonal is a suborbifold.

*Remark 39.* Suborbifolds are therefore orbifolds whose orbifold structure is a "substructure" of the ambient orbifold $\mathcal{O}$, in the sense that the restriction to $\mathcal{P}$ of an orbifold map from $\mathcal{O}$ are themselves orbifold maps from $\mathcal{P}$.

*Remark 40.* Let $\mathcal{P} \subset \mathcal{O}$ be a suborbifold. Note that even though a point $p \in X_{\mathcal{P}}$ may be in the singular set of $\mathcal{O}$, it need not be in the singular set of $\mathcal{P}$.

**Example 41.** Let $\mathcal{O}$ be an orbifold and $\mathcal{O} \times \mathcal{O}$ be the orbifold product of $\mathcal{O}$ with itself. Let $\Delta : \mathcal{O} \to \mathcal{O} \times \mathcal{O}$ be the diagonal mapping. By construction, the *diagonal* $= \Delta(\mathcal{O}) \subset \mathcal{O} \times \mathcal{O}$ is a suborbifold of $\mathcal{O} \times \mathcal{O}$ with isotropy group $\Gamma_{(x,x)} \cong \Gamma_x$ via the diagonal action $\gamma \cdot (\tilde{x}, \tilde{x}) = (\gamma \cdot \tilde{x}, \gamma \cdot \tilde{x})$.

**Example 42.** Let $f \in \text{Diff}_{\text{Orb}}^r(\mathcal{O})$, then the graph of $f$, graph$(f)$ defined by

$$\text{graph}(f) = \{(x,y) \in \mathcal{O} \times \mathcal{O} \mid y = f(x)\}$$

in $\mathcal{O} \times \mathcal{O}$ is a suborbifold of $\mathcal{O}$. Note the isotropy group $\Gamma_{(x,y)} \cong \Gamma_x$ acting via the twisted diagonal action $\gamma \cdot (\tilde{x}, \tilde{y}) = (\gamma \cdot \tilde{x}, \Theta_{f,x}(\gamma) \cdot \tilde{y})$. Similarly given an orbisection $\sigma$ of an orbibundle $\mathcal{P} \to \mathcal{O}$, this defines a suborbifold of the total space of the orbibundle $\mathcal{P}$.

**Definition 43.** Let $\mathcal{P}$ be an $m$-dimensional $C^r$ suborbifold of an $n$-dimensional $C^r$ orbifold $\mathcal{O}$ (where $r \geq 1$). The *normal orbibundle* $N\mathcal{P}$ of

$\mathcal{P}$ in $\mathcal{O}$ is the linear orbibundle over $\mathcal{P}$ with projection $p : N\mathcal{P} \to \mathcal{P}$ so that if $U$ is an orbifold chart in $\mathcal{P}$ about $x \in \mathcal{P}$ then

$$\tilde{p}^{-1}(\tilde{U}) = \tilde{U} \times (\mathbb{R}^n/\mathbb{R}^m)$$

with the $\Gamma_x$ action defined by

$$\gamma \cdot (x, v/\mathbb{R}^m) = (\gamma \cdot x, d\gamma_x(v/\mathbb{R}^m)).$$

## 6  Extending Orbifold Diffeomorphisms and Local Contractions

For any subgroup $G$ of the homeomorphism group $H(X)$ of a topological space $X$, let $G_c \subset G$ denote those elements of $G$ with compact support in $X$. Let $G_0$ be the subgroup of $G_c$ whose elements are isotopic to the identity through elements of $G$ with compactly supported isotopy. For any self-map $f : X \to X$ of a topological space $X$, let the support $\mathrm{supp}(f) = \mathrm{cl}\{x \in X \mid f(x) \neq x\}$ where $\mathrm{cl}(S)$ denotes the closure of the set $S$. By compactly supported isotopy we mean an isotopy $f : [0,1] \times X$, such that $\mathrm{supp}(f) \subset [0,1] \times X$ is compact. The proofs of the following results can be found in Borzellino and Brunsden.[10]

**Proposition 44.** *Let $\Sigma$ denote the singular set of an orbifold $\mathcal{O}$. The group $\mathrm{Diff}^0(\mathcal{O} - \Sigma)_c$ is a subgroup of $\mathrm{Diff}^0_{\mathrm{red}}(\mathcal{O})$ for any topological orbifold $\mathcal{O}$. Moreover, if $\mathcal{O}$ is $C^r$-smooth, then for each component $\mathcal{A} = \Sigma_m^{(l_1)(l_2)\cdots(l_m)}$ of $\Sigma_m$ and $f \in \mathrm{Diff}^r(\mathcal{A} - \Sigma_{\mathcal{A}})_0$, there is an extension $g \in \mathrm{Diff}^0_{\mathrm{red}}(\mathcal{O})$ and a neighborhood $U$ of $\Sigma_m^{(l_1)(l_2)\cdots(l_m)}$ in $\mathcal{O}$ such that $\mathrm{supp}(g) \subset U$ and the restriction $g|_{\Sigma_m^{(l_1)(l_2)\cdots(l_m)}} = f$ for any $1 \leq r \leq \infty$.*

The proof Theorems A and Theorem B will require that there are enough local orbifold diffeomorphisms whose behavior under the group isomorphism can be controlled. To this end, we use the following

**Definition 45.** For a locally compact Hausdorff space $X$, a subgroup $G \subset H(X)$ and $x \in X$, we say that $g_x \in G$ is a *local contraction about $x$* if:

1. $x \in \mathrm{supp}(g_x)$ and $\mathrm{supp}(g_x)$ is compact

2. for all open neighborhoods $V$ and $W$ of $x$ in $\mathrm{supp}(g_x)$ with $\overline{W} \subset V \subset \overline{V} \subset \mathrm{int}(\mathrm{supp}(g_x))$ there is an $N \in \mathbb{N}$ so that $g_x^n(\overline{V}) \subset W$ for all $n > N$.

3. $g_x(x) = x$

For locally smooth orbifolds there are plenty of local contractions:

**Proposition 46.** *If $\mathcal{O}$ is locally smooth, then for each $x \in \mathcal{O}$ and neighborhood $U$ of $x$ there is a (reduced) local contraction about $x$ with support in $U$.*

# 7    Proof of Theorem A

We sketch an outline of the proof of Theorem A in the locally smooth case. The full proof, including the extension to the smooth case may be found in Borzellino and Brunsden.[10] Note that for any open subset $U$ of an orbifold $\mathcal{O}$, $x \in U$ if and only if there is a neighborhood $V$ of $x$ so that $V - \Sigma \subset U - \Sigma$ and for an open subset $U$ as above, $x \in \mathrm{cl}(U)$ if and only if $(V - \Sigma) \cap (U - \Sigma)$ $\neq \emptyset$ for all neighborhoods $V$ of $x$. Note that these follow almost trivially from the nowhere density of the singular set. In brief, the outline of the proof is as follows.

Let $\mathcal{O}_1$ and $\mathcal{O}_2$ be two compact, locally smooth orbifolds and let $\Phi :$ $\mathrm{Diff}^r_{\mathrm{red}}(\mathcal{O}_1) \to \mathrm{Diff}^r_{\mathrm{red}}(\mathcal{O}_2)$ be a group isomorphism. By Lemmas 25 and 26, Proposition 44 and Rubin's theorem we have a homeomorphism $h : \mathcal{O}_1 - \Sigma_1 \to$ $\mathcal{O}_2 - \Sigma_2$ such that for every $f \in \mathrm{Diff}^r_{\mathrm{red}}(\mathcal{O}_1)$ we have $\Phi(f) = hfh^{-1}$. Note that this implies that the singular sets of $\mathcal{O}_i$ are either both empty or are both non-empty. To see this, suppose $\Sigma_1 \neq \emptyset$ and that $\Sigma_2 = \emptyset$. Then $\mathcal{O}_2 = \mathcal{O}_2 - \Sigma_2$ is a closed manifold. $\mathcal{O}_1 - \Sigma_1$, however, is a non–compact manifold, and this contradicts the existence of a homeomorphism $h : \mathcal{O}_1 - \Sigma_1 \to \mathcal{O}_2 - \Sigma_2$ guaranteed by Rubin's theorem. Since Rubin's theorem implies Theorem A when $\Sigma_1 = \Sigma_2 = \emptyset$ (the manifold case), we need only concern ourselves with case when $\Sigma_1$ and $\Sigma_2$ are non–empty.

Next, we extend $h$ to a bijection $\overline{h} : \mathcal{O}_1 \to \mathcal{O}_2$ inducing the group isomorphism as follows:

Let $x \in \Sigma_1$, and let $U_x$ be a relatively compact open neighborhood of $x$ in $\mathcal{O}_1$. By Proposition 46, there exists a $g_x \in \mathrm{Diff}^r_{\mathrm{red}}(\mathcal{O})$ which is a local contraction about $x$ with support in $U_x$. Let $\hat{g}_x = \Phi(g_x)$, and $\hat{U}_x = \mathrm{int}\left(\mathrm{cl}(h(U_x - \Sigma_1))\right)$. It follows from Rubin's theorem that $\mathrm{supp}(\hat{g}_x) \subset \mathrm{cl}(\hat{U}_x)$. For details, see our paper.[10]

We now show that $\hat{g}_x$ possesses a non-empty invariant set $Y_x \subset$ $\mathrm{int}(\mathrm{supp}(\hat{g}_x)) \cap \Sigma_2$. For this, Let $\hat{W} \subset \hat{U}_x$ be any relatively compact open subset of $\mathcal{O}_2$ with

$$x \in \mathrm{int}\left(\mathrm{cl}(h^{-1}(\hat{W} - \Sigma_2))\right)$$

Compactness of $\mathcal{O}_2$ makes this possible. For any neighborhood $V$ of $x$ with $\mathrm{cl}(V) - \Sigma_1 \subset h^{-1}(\hat{W} - \Sigma_2)$ there is an $m > 0$ so that

$$g_x^m\left(h^{-1}(\hat{W} - \Sigma_2)\right) \subset V \subset \mathrm{int}\left(\mathrm{cl}(h^{-1}(\hat{W} - \Sigma_2))\right)$$

since $g_x$ is a local contraction about $x$. Therefore,

$$x \in \bigcap_{n < N} g_x^{mn}\left(\mathrm{cl}\left(h^{-1}(\hat{W} - \Sigma_2)\right)\right) \neq \emptyset$$

which implies

$$\bigcap_{n<N} \mathrm{cl}\left(g_x^{mn}\left(h^{-1}(\hat{W} - \Sigma_2)\right)\right) \neq \emptyset$$

and so by definition of $\hat{g}_x$ and $h$,

$$\bigcap_{n<N} \mathrm{cl}\left(h^{-1}\left(\hat{g}_x^{mn}(\hat{W} - \Sigma_2)\right)\right) \neq \emptyset$$

which in turn implies,

$$\bigcap_{n<N} h^{-1}\left(\hat{g}_x^{mn}(\hat{W}) - \Sigma_2\right) \neq \emptyset$$

It now follows that

$$\emptyset \neq \bigcap_{n<N} h \circ h^{-1}\left(\hat{g}_x^{mn}(\hat{W}) - \Sigma_2\right) \subset \bigcap_{n<N} \hat{g}_x^{mn}(\hat{W})$$

and so

$$\bigcap_{n<N} \hat{g}_x^{mn}(\mathrm{cl}(\hat{W})) \neq \emptyset$$

Then the collection of closed sets $\{\hat{g}_x^{mn}(\mathrm{cl}(\hat{W}))\}$ has the finite intersection property, and so by compactness of $\mathcal{O}_2$ we have

$$Y_x = \bigcap_{n>0} \hat{g}_x^{mn}(\mathrm{cl}(\hat{W})) \neq \emptyset.$$

By construction, $Y_x = \bigcap_{m>0} \hat{g}_x^{mn}(\mathrm{cl}(\hat{W}))$ is a compact, $\hat{g}_x$ invariant set. We claim that $Y_x$ is independent of $g_x$ and the subset $\hat{W}$. To see this, suppose that $g_x'$ is another local contraction with fixed point $x$, and $\hat{W}' \subset \mathcal{O}_2$ is a compact subset of $\mathrm{int}\left(\mathrm{supp}(\Phi(g_x'))\right)$ satisfying the same requirement of $\hat{W}$ as above. As both $g_x$ and $g_x'$ are local contractions, for any $n > 0$ there is an $m > 0$ so that:

$$g_x^m\left(\mathrm{int}(\mathrm{cl}(h^{-1}(\hat{W} - \Sigma_2)))\right) \subset g_x'^n\left(\mathrm{int}(\mathrm{cl}(h^{-1}(\hat{W}' - \Sigma_2)))\right)$$

and for any $m > 0$ there is an $n > 0$ so that:

$$g_x'^n\left(\mathrm{int}(\mathrm{cl}(h^{-1}(\hat{W}' - \Sigma_2)))\right) \subset g_x^m\left(\mathrm{int}(\mathrm{cl}(h^{-1}(\hat{W} - \Sigma_2)))\right)$$

Therefore $\bigcap_{n>0} \hat{g}_x^n(\hat{W}) \subset \bigcap_{m>0} \hat{g}_x'^m(\hat{W}') \subset \bigcap_{n>0} \hat{g}_x^n(\hat{W})$ which shows the independence of $Y_x$ on the local contraction.

Let $g_x$ and $g_{x'}'$ be local contractions about $x$ and $x'$ respectively with disjoint supports such that $\mathrm{supp}(g_x) \subset U$ and $\mathrm{supp}(g_{x'}') \subset U'$ where $U$ and $U'$

are open sets with $U \cap U' = \emptyset$, $U = \text{int}(\text{cl}(U))$ and $U' = \text{int}(\text{cl}(U'))$. Therefore $h(U - \Sigma_1) \cap h(U' - \Sigma_1) = \emptyset$ and by the remark above, if $z \in \text{int}\left(\text{cl}(h(U - \Sigma_1))\right)$, then there is a neighborhood $V$ of $z$ so that $V - \Sigma_2 \subset \text{int}\left(\text{cl}(h(U - \Sigma_1))\right) - \Sigma_2 = h(U - \Sigma_1)$. Therefore $z \notin \text{int}\left(\text{cl}(h(U' - \Sigma_1))\right)$. Reversing the roles of $U$ and $U'$ shows that

$$\text{int}\left(\text{cl}(h(U - \Sigma_1))\right) \bigcap \text{int}\left(\text{cl}(h(U' - \Sigma_1))\right) = \emptyset$$

Since $Y_x \subset \text{int}\left(\text{cl}(h(U - \Sigma_1))\right)$ and $Y_{x'} \subset \text{int}\left(\text{cl}(h(U' - \Sigma_1))\right)$, $Y_x \cap Y_{x'} = \emptyset$. Therefore for any two such subsets $Y_x$ and $Y_{x'}$ of $\mathcal{O}_2$, if $Y_x \cap Y_{x'} \neq \emptyset$ then $Y_x = Y_{x'}$ and $x = x'$.

Given a $k \in \text{Diff}^r_{\text{red}}(\mathcal{O}_1)$, $x \in \Sigma_1$ and a local contraction $g_x$ about $x$, the orbifold diffeomorphism $k \circ g_x \circ k^{-1}$ is a local contraction about $k(x)$. Hence $\Phi(k \circ g_x \circ k^{-1})$ will have invariant set $Y_{k(x)}$. Since $\Phi$ is a group isomorphism between $\text{Diff}^r_{\text{red}}(\mathcal{O}_1)$ and $\text{Diff}^r_{\text{red}}(\mathcal{O}_2)$, the invariant set of $\Phi(k \circ g_x \circ k^{-1})$ will be $\Phi(k)(Y_x)$. Therefore $\Phi(k)(Y_x) = Y_{k(x)}$ for all $x \in \Sigma_1$. We will use this below to prove that the sets $Y_x$ consist of a single point.

Let $y \in Y_x$, and $\hat{g}_y \in \text{Diff}^r_{\text{red}}(\mathcal{O}_2)$ be a local contraction about $y$. Let $g_y = \Phi^{-1}(\hat{g}_y)$ and then by definition $y \in \hat{g}_y^n(Y_x) = Y_{g_y^n(x)}$ for all $n \geq 0$. Hence $Y_x \cap \hat{g}_y^n(Y_x) \neq \emptyset$ for all $n \geq 0$ and so $Y_x = \hat{g}_y^n(Y_x)$ for all $n \geq 0$. If $z \in Y_x \cap \text{supp}(\hat{g}_y)$ then for any neighborhood $V$ of $y$ in $\mathcal{O}_2$, there is an $n > 0$ so that $\hat{g}_y^n(z) \in V$ which implies that $Y_x \cap \text{int}(\text{supp}(\hat{g}_y)) = \{y\}$. Since $\hat{g}_y$ was essentially arbitrary, this implies that $Y_x = \{y\}$, that is, the invariant set $Y_x$ of $g_x$ consists of a single point.

Define the extension $\overline{h}$ of $h$ to all of $\mathcal{O}_1$ by the following:

$$\overline{h}(x) = \begin{cases} h(x), & \text{if } x \in \mathcal{O}_1 - \Sigma_1 \\ Y_x, & \text{if } x \in \Sigma_1 \end{cases}$$

By construction, $\overline{h}$ is an injection inducing the group isomorphism. Similarly we can construct an injection $\overline{h}^{-1}$. Continuity of $\overline{h}$ follows from the following. Given $x \in \mathcal{O}_1$ and a neighborhood $U_x$ of $x$, then there is a local contraction $g_x$ about $x$ with support in $U_x$ (by Proposition 46). By construction, $x \in \text{int}(\text{supp}(g_x))$ and so the collection

$$\mathcal{B} = \bigcup_{x \in \mathcal{O}_1} \bigcup_{U_x \ni x} \{\text{int}(\text{supp}(g_x)) \mid \text{int}(\text{supp}(g_x)) \subset U_x\}$$

forms a base for the topology of $\mathcal{O}_1$. Let $\text{Fix}(f) = \{x \in \mathcal{O} \mid f(x) = x\}$. Thus

$$\overline{h}\left((\mathcal{O}_1 - \text{int}(\text{supp}(g_x))) \cup \{x\}\right) = \overline{h}\left(\text{Fix}(g_x)\right) = \text{Fix}(\Phi(g_x))$$

$$= (\mathcal{O}_2 - \text{int}(\text{supp}(\Phi(g_x)))) \cup \{\overline{h}(x)\}$$

so

$$\overline{h}(\mathcal{O}_1 - \text{int}(\text{supp}(g_x))) = \mathcal{O}_2 - \text{int}\left(\text{supp}(\Phi(g_x))\right)$$

and therefore

$$\overline{h}\left(\text{int}(\text{supp}(g_x))\right) = \text{int}\left(\text{supp}(\Phi(g_x))\right)$$

and so $\overline{h}$ maps basic open sets to basic open sets and so $\overline{h}$ is continuous. Similarly, $\overline{h^{-1}}$ is continuous. Note that by construction

$$\overline{h} \circ \overline{h^{-1}} = \text{Id on } \mathcal{O}_2 - \Sigma_2$$

and

$$\overline{h^{-1}} \circ \overline{h} = \text{Id on } \mathcal{O}_1 - \Sigma_1.$$

Since $\mathcal{O}_2 - \Sigma_2$ is dense in $\mathcal{O}_2$ and $\mathcal{O}_1 - \Sigma_1$ is dense in $\mathcal{O}_1$, we have that $\overline{h} \circ \overline{h^{-1}} = \text{Id}$ on $\mathcal{O}_2$ and $\overline{h^{-1}} \circ \overline{h} = \text{Id}$ on $\mathcal{O}_1$. Hence $\overline{h^{-1}} = (\overline{h})^{-1}$ and so $\overline{h}$ is a homeomorphism that induces the group isomorphism $\Phi$. The proof of the main part of Theorem A is complete.

## 8 Proof of Theorem B

To prove Theorem B, given a group isomorphism $\tilde{\Phi} : \text{Diff}^r_{\text{Orb}}(\mathcal{O}_1) \rightarrow \text{Diff}^r_{\text{Orb}}(\mathcal{O}_2)$, we first show that $\tilde{\Phi}$ induces a group isomorphism $\Phi : \text{Diff}^r_{\text{red}}(\mathcal{O}_1) \rightarrow \text{Diff}^r_{\text{red}}(\mathcal{O}_2)$. Let $\tau : \text{Diff}^r_{\text{Orb}}(\mathcal{O}) \rightarrow \text{Diff}^r_{\text{red}}(\mathcal{O})$ denote the forgetful homomorphism. Define $\Phi$ by $\Phi(f) = \tau \circ \tilde{\Phi} \circ \tau^{-1}$. We show that this map is well–defined. Suppose $\tilde{f}_1, \tilde{f}_2 \in \tau^{-1}(f)$. Note that $\tilde{f}_1|_{\mathcal{O}_1 - \Sigma_1} = \tilde{f}_2|_{\mathcal{O}_1 - \Sigma_1}$. If we consider the manifolds $\mathcal{O}_i - \Sigma_i$, the groups of diffeomorphisms $\text{Diff}^r_{\text{Orb}}(\mathcal{O}_i)|_{\mathcal{O}_i - \Sigma_i}$, (the restrictions of elements of $\text{Diff}^r_{\text{Orb}}(\mathcal{O}_i)$ to $\mathcal{O}_i - \Sigma_i$), and the group isomorphism $\tilde{\Phi}$ between them, the hypotheses of Rubin's theorem are satisfied (by Lemma 26), and thus there exists a homeomorphism $h : \mathcal{O}_1 - \Sigma_1 \rightarrow \mathcal{O}_2 - \Sigma_2$ with $\tilde{\Phi}(\tilde{f}) = h \circ \tilde{f} \circ h^{-1}$ for all $\tilde{f} \in \text{Diff}^r_{\text{Orb}}(\mathcal{O}_1)|_{\mathcal{O}_1 - \Sigma_1}$. We then conclude that $\tilde{\Phi}(\tilde{f}_1) = \tilde{\Phi}(\tilde{f}_2)$ on $\mathcal{O}_2 - \Sigma_2$. This is enough to conclude that $\tau \circ \tilde{\Phi}(\tilde{f}_1) = \tau \circ \tilde{\Phi}(\tilde{f}_2)$ as elements of $\text{Diff}^r_{\text{red}}(\mathcal{O}_2)$.

By Theorem A, there is a homeomorphism $h : X_{\mathcal{O}_1} \rightarrow X_{\mathcal{O}_2}$ inducing $\Phi$. The remainder of the proof is to show that $h$ has local lifts in neighborhoods of each $x \in \Sigma_1$ that are $\Gamma_x$ equivariant once there are the appropriate homomorphisms between $x$ and $h(x)$. To show that $h$ has the appropriate local lifting properties we examine the group $S(\mathcal{O}) = \ker(\tau : \text{Diff}^r_{\text{Orb}}(\mathcal{O}) \rightarrow \text{Diff}^r_{\text{red}}(\mathcal{O}))$. Note that it is enough to examine those automorphisms that cover the identity. The previous shows that since $\Phi(\text{Diff}^r_{\text{red}}(\mathcal{O}_1)) = \text{Diff}^r_{\text{red}}(\mathcal{O}_2)$, it follows that $\tilde{\Phi}(S(\mathcal{O}_1)) = S(\mathcal{O}_2)$. The homomorphism $\tilde{\Phi}$ induces a local homomorphism $\Theta_x : \Gamma_x \to \Gamma_{h(x)}$ of isotropy groups as follows. Let $(\tilde{U}_x, \Gamma_x, \rho_x, \phi_x)$ be an orbifold chart around $x \in \Sigma_1$, $y = h(x)$ and $(\tilde{U}_y, \Gamma_y, \rho_y, \phi_y)$ an orbifold chart around $y$ so that

$$h(\phi_x(\tilde{U}_x/\Gamma_x)) = \phi_y(\tilde{U}_y/\Gamma_y)$$

Let $D_x \subset \tilde{U}_x$ and $D'_y \subset \tilde{U}_y$ be Dirichlet domains for the actions of $\Gamma_x$ and $\Gamma_y$ respectively. For $z \in D_x$ there is a unique $\gamma \in \Gamma_x$ so that $\gamma^{-1} \cdot z \in D_x$. Let $\sigma_\gamma \in S(\mathcal{O}_1)$ be the element that sends $\tilde{U}_x \ni z \to \gamma \cdot z$ (See Proposition 28). Define $\Theta_x : \Gamma_x \to \Gamma_y$ by $\Theta_x(\gamma) = \gamma'$ where $\tilde{\Phi}(\sigma_\gamma) = \sigma'_{\gamma'} \in S(\mathcal{O}_2)$. It is easily checked that this is a group homomorphism. For any $z \in \tilde{U}_x$, we define a $\Theta_x$ equivariant lift $\bar{h}$ of $h$ by first defining it for $z \in D_x$ as the unique $\bar{h}(z) = z' \in D'_y$ so that

$$\phi_y(\Gamma_y \cdot z') = h(\phi_x(\Gamma_x \cdot z))$$

and then extending to $z \in \gamma \cdot D_x$ by letting $\bar{h}(z) = z'$ where

$$\Theta_x(\gamma^{-1}) \cdot z' \in D'_y$$

and

$$\phi_y(\Gamma_y \cdot z') = h(\phi_x(\Gamma_x \cdot z))$$

By construction, this is a $\Theta_x$ equivariant lift of $h$ to a neighborhood of $x$ and so the triple $(h, \Theta_x, \bar{h})$ is an (unreduced) orbifold homeomorphism inducing the group isomorphism $\tilde{\Phi}$.

## 9  Proof of Theorem C

We sketch the proof of this result. Complete proofs may be found in Borzellino and Brunsden.[12,13] The general idea is to mimic the details as much as possible of the analogous proof for manifolds. That this is possible leads to some extensions of several results on group actions.

It suffices to construct a neighborhood of the identity homeomorphic to some open set of the appropriate topological vector space (a Banach space for $1 \le r < \infty$ and a Fréchet space for $r = \infty$). Let $\mathcal{NO}$ be the normal orbibundle of the diagonal $\Delta(\mathcal{O})$ in $\mathcal{O} \times \mathcal{O}$. That is, it is locally the quotient of $T_{\Delta(\mathcal{O})}(\mathcal{O} \times \mathcal{O})/\tilde{T}\Delta(\mathcal{O})$, i.e. the normal orbibundle of the diagonal in the Cartesian product $\mathcal{O} \times \mathcal{O}$. There is an orbifold tubular neighborhood of $\Delta(\mathcal{O})$ that is covered by $\mathcal{NO}$. With a little work, one can see that $\mathcal{NO}$ is isomorphic to $T\mathcal{O}$. We let

$$\overline{\exp} : \mathcal{NO} \to \mathcal{O} \times \mathcal{O}$$

be the exponential map induced by a Riemannian metric on $\mathcal{O}$. This induces a map which by abuse of notation we also call $\overline{\exp}$,

$$\overline{\exp} : D^r_{\mathcal{O}}(\mathcal{O}) \to C^r_{\text{Orb}}(\mathcal{O}, \mathcal{O})$$

Where $D^r_{\mathcal{O}}(\mathcal{O})$ is the topological vector space of $C^r$ orbisections of the orbibundle $\mathcal{NO}$. The above defines a $C^0$ neighborhood of the identity (which corresponds to $\overline{\exp}(0)$) as in the manifold case. A sufficiently small $C^1$ neighborhood of the identity in $C^r_{\text{Orb}}(\mathcal{O}, \mathcal{O})$ is in $\text{Diff}^r_{\text{Orb}}(\mathcal{O})$ and so this gives $\text{Diff}^r_{\text{Orb}}(\mathcal{O})$ a manifold structure. For details, see Borzellino and Brunsden.[13]

## Acknowledgements

The authors would like to thank the organizers and Howard University for their generosity and hospitality. We also wish to thank the other participants for a most interesting and stimulating conference.

## References

1. M. Anderson and J. Cheeger, *Diffeomorphism finiteness for manifolds with Ricci curvature and $L^{n/2}$-norm of curvature bounded*, Geom. Funct. Anal. **1**, no. 3, 231–252 (1991).
2. A. Banyaga, *On Isomorphic Classical Diffeomorphism Groups I*, Proc. Amer. Math. Soc. **98**, 113–118 (1986).
3. A. Banyaga, *On Isomorphic Classical Diffeomorphism Groups II*, J. Diff. Geometry **28**, 23–25 (1988).
4. A. Banyaga, *The structure of classical diffeomorphism groups*, Mathematics and its Applications, vol. 400, Kluwer Academic Publishers, Dordrecht, 1997.
5. A. Banyaga, *Isomorphisms Between Classical Diffeomorphism Groups*, CRM Proceedings and Lecture Notes **15**, 1–15 (1998).
6. A. Banyaga, *Sur la Structure du Groupe des Difféomorphismes qui Préservent une Forme Symplectique*, Comment. Math. Helv. **53**, no. 2, 174–227 (1998).
7. J. Borzellino, *Riemannian Geometry of Orbifolds*, PhD. Thesis, Univ. Calif. Los Angeles, Spring 1992.
8. J. Borzellino. *Orbifolds of Maximal Diameter*, Indiana U. Math. J. **42**, 37–53 (1993).
9. J. Borzellino. *Orbifolds With Lower Ricci Curvature Bounds*, Proc. American Mathematical Society **125**, no. 10, 3011–3018 (1997).

10. J. Borzellino and V. Brunsden, *Determination of the topological structure of an orbifold by its group of orbifold diffeomorphisms*, Preprint (1999).
11. J. Borzellino and V. Brunsden, *An Automorphism of the Orbifold Diffeomorphism Group which is not an Inner Automorphism*, Preprint (2000).
12. J. Borzellino and V. Brunsden, *A Differential Topologist's Toolkit for Orbifolds*, In preparation (2002).
13. J. Borzellino and V. Brunsden, *On the Topology of the Group of Orbifold Diffeomorphisms of a Smooth Orbifold*, In preparation (2002).
14. J. Borzellino and S. Zhu. *The Splitting Theorem for Orbifolds*, Illinois J. Math. **38**, 679–691 (1994).
15. G. Bredon, *Introduction to the Theory of Transformation Groups*, Academic Press, New York 1972.
16. V. Brunsden, *Local Rigidity and Group Cohomology I: Stowe's Theorem for Banach Manifolds*, Bull. Aus. Math. Soc. **59**, no. 2, 271–295 (1999).
17. V. Brunsden, *Local Rigidity and Group Cohomology II: Anosov-like Actions*, preprint (2000).
18. A. Dress, *Newman's Theorems on Transformation Groups*, Topology **8**, 203–207 (1969).
19. R. Filipkiewicz, *Isomorphisms between diffeomorphism groups*, Ergodic Theory Dynam. Systems **2**, 159–171 (1982).
20. K. Fukaya and K. Ono, *Arnold conjecture and Gromov-Witten invariant*, Topology **38**, no. 5, 933–1048 (1999).
21. M. Hirsch, *Differential Topology*, Springer-Verlag, New York, 1976.
22. M. Rubin, *On the reconstruction of topological spaces from their homeomorphism groups*, Trans. Amer. Math. Soc. **312**, no. 2, 437–538 (1989).
23. T. Rybicki, *Isomorphisms between groups of diffeomorphisms*, Proc. Amer. Math. Soc. **123**, no. 1, 303–310 (1995).
24. I. Satake. *On a Generalization of the Notion of Manifold*, Proc. Nat. Acad. Sci. USA **42**, 359–363 (1956).
25. I. Satake. *The Gauss–Bonnet Theorem for V–Manifolds*, Jour. Math. Soc. Japan **9**, 464–492 (1957).
26. P. Scott. *The Geometries of 3–manifolds*, Bull. London Math Soc. **15**, 401–487 (1983).
27. W. Thurston, *The Geometry and Topology of 3-Manifolds*, Lecture notes, Princeton University Mathematics Department, 1978.

# A NOTE ON ISOTOPIES OF SYMPLECTIC AND POISSON STRUCTURES

AUGUSTIN BANYAGA

*Department of Mathematics,*
*The Pennsylvania State University, University Park, PA 16802 ,*
*E-Mail: banyaga@math.psu.edu*

PAUL DONATO

*U.R.A. 225 (C.N.R.S.),*
*Université de Provence, France,*
*E-Mail: donato@gyptis.univ-mrs.fr*

We give sufficient conditions for two symplectic forms to be isotopic in terms of the corresponding Poisson structures. We deduce a Moser-type theorem for the images under the musical isomorphism of harmonic 2-forms on compact symmetric Riemannian manifolds. Finally, we give a theorem on smooth liftings of Poisson-Lichnerowicz coboundaries

## 1    Preliminaries [7]

A symplectic form on a 2n dimensional manifold $M$ is a closed 2-form $\Omega$ such that the bundle map $\tilde{\Omega}: T(M) \to T^*(M)$ defined by $\tilde{\Omega}(X)(\xi) = \Omega(X, \xi)$ for all vector fields $X$ and $\xi$, is an isomorphism. The inverse map $(\tilde{\Omega})^{-1}: T^*(M) \to T(M)$ defines a Poisson structure.

A Poisson structure on a smooth manifold $M$ is a bivector $\pi$ ( i.e. a section of $\Lambda^2 T(M)$) such that $[\pi, \pi] = 0$, where $[.]$ is the Schouten bracket (this is a natural extension of the Lie derivative to skew-symmetric contravariant tensor fields). The Poisson structure defines a bundle map $\tilde{\pi}: T^*(M) \to T(M)$ : $\beta(\tilde{\pi}(\alpha)) = \pi(\alpha, \beta)$ for all 1-forms $\alpha$ and $\beta$. If $\tilde{\pi}$ is invertible, we get a symplectic form $\Omega$ defined as

$$\Omega(X, Y) = \pi((\tilde{\pi})^{-1}(X), (\tilde{\pi})^{-1}(Y)) \text{ i.e } (\tilde{\Omega})^{-1} = \tilde{\pi}.$$

The space of $k$-vectors, i.e. smooth sections of $\Lambda^k T(M)$, will be denoted by $\mathcal{V}^k(M)$.

The map

$$\delta: \mathcal{V}^p(M) \to \mathcal{V}^{p+1}(M), \qquad u \mapsto [\pi, u]$$

on the spaces $\mathcal{V}^*(M)$ of multivectors, satisfies $\delta^2 = 0$, hence $(\mathcal{V}^*(M), \delta)$ is a cochain complex and its cohomology $H^*_\pi(M)$ , introduced by Lichnerowicz

[4], is called the Lichnerowicz-Poisson cohomology of the Poisson manifold $(M, \pi)$. The bundle map $\tilde{\pi}$ induces a homomorphism

$$\pi_*: H^*(M, \mathbf{R}) \to \mathbf{H}_\pi^*(\mathbf{M})$$

from the de Rham cohomology to the Lichnerowicz-Poisson cohomology. This homomorphism is an isomorphism if $\pi$ comes from a symplectic structure.

## 2 The main results

Two symplectic forms $\Omega$, $\Omega'$ on a smooth manifold $M$ are said to be isotopic if they represent the same cohomology class $a$ in $H^2(M, \mathbf{R})$, and there exists a smooth family of symplectic forms $\Omega_t$, all in the class $a$, and such that $\Omega_0 = \Omega$ and $\Omega_1 = \Omega'$.

The well known theorem of Moser [5] asserts that two symplectic forms $\Omega$ and $\Omega'$ on a compact manifold $M$ are isotopic iff there exists a diffeomorphism $\phi$ of $M$, isotopic to the identity such that $\phi^*\Omega = \Omega'$.

This beautiful classification theorem for symplectic structures has a weakness: it is a difficult problem to decide whether two symplectic forms are isotopic.

In this note, we give a simple characterization of isotopic symplectic structures in terms of the associated Poisson structures.

**Theorem 1** *Let $\Omega$ and $\Omega'$ be two symplectic forms on a compact manifold $M$. Let $\pi$ and $\pi'$ be the corresponding Poisson structures. Suppose that $[\pi, \pi'] = 0$ and that there exists a vector field $X$ such that $\pi' = \pi + L_X \pi$ and $L_X(L_X \pi) = 0$, then $\Omega$ and $\Omega'$ are isotopic. Here $L_X \eta = [\eta, X]$ is the Lie derivative of the p-vector $\eta$ in the direction $X$.*

The discussions in this note lead to the following classification of bivectors dual to harmonic 2-forms on symmetric Riemannian manifolds:

**Theorem 2** *Let $\omega$ and $\omega'$ be 2 harmonic 2-forms on a compact symmetric Riemannian manifold $(M, g)$. Consider the bivectors $\pi$, $\pi'$ corresponding to $\omega$, $\omega'$ by the musical isomorphism of the Riemannian metric $g$. If there exists a vector field $X$ such that $\pi' = \pi + L_X \pi$ and $L_X(L_X \pi) = 0$, then there exists a diffeomorphism $\phi$ isotopic to the identity such that $\phi_* \pi = \pi'$.*

## 3  Isotopies of symplectic and Poisson structures

Let $(M, \pi)$ be a Poisson manifold. The condition $[\pi, \pi] = \delta\pi = 0$ means that $\pi$ represents a class $a \in H^2_\pi(M)$. Any other Poisson structure $\pi'$ represents an element in $H^2_\pi(M)$ if $[\pi, \pi'] = 0$, i.e. if $\pi$ and $\pi'$ are compatible.

Clearly the condition $[\pi, \pi'] = 0$ implies that $\pi_t = t\pi' + (1-t)\pi$ is a smooth family of compatible Poisson structures. If $\pi'$ represents the same class as $\pi$ in $H^2_\pi(M)$, then, the deformation above stays in the same cohomology class.

Moreover, $\pi$ and $\pi'$ are in the same class iff there exists a vector field $X$ such that $\pi' = \pi + [\pi, X]$. Recall that $[\pi, X] = L_X\pi$. The deformation above can also be written : $\pi_t = \pi + [\pi, X_t]$ with $X_t = tX$.

Suppose now the Poisson structures $\pi$ and $\pi'$ come from two arbitrary symplectic forms $\Omega$ and $\Omega'$, we have an isomorphism $a^* : H^*_\pi(M) \to H^*_{\pi'}(M)$ which is the composition of the following isomorphisms:

$$H^*_\pi(M) \approx H^*(M, \mathbf{R}) \approx \mathbf{H}^*_{\pi'}(\mathbf{M}).$$

This isomorphism takes the class of $\pi$ in $H^2_\pi(M)$ to the class of $\pi'$ in $H^2_{\pi'}(M)$. However $\pi'$ represents a class in $H^2_\pi(M)$ only if $[\pi, \pi'] = 0$, i.e. if $\pi$ and $\pi'$ are compatible. Moreover, the bivectors $\pi$ and $\pi'$ represent the same class in $H^2_\pi(M)$
iff $\pi' = \pi + [\pi, X]$ for some vector field $X$. Note that $X$ is not unique, since we can add to it any automorphism of $\pi$, i.e any vector filed $Y$ such that $L_Y\pi = [\pi, Y] = 0$.

**Lemma 1** *Let $\pi, \pi'$ be two compatible Poisson structures on a compact manifold $M$. Suppose there exists a vector field $X$ such that $\pi' = \pi + L_X\pi$ and $L_X(L_X\pi) = 0$, then there exists a diffeomorphism $\phi$ isotopic to the identity such that $\phi_*\pi = \pi'$*

**Proof** Consider the deformation $\pi_t = \pi + [\pi, X_t]$ with $X_t = tX$. Let $Y_t = -X$, then: $L_{Y_t}(\pi_t) = -[\pi_t, X] = -([\pi, X] + t[[\pi, X], X]) = -[\pi, X]$ since $[[\pi, X], X] = 0$ by assumption. Now $\partial/\partial t(\pi_t) = [\pi, X]$, therefore , we have:

$$L_{Y_t}\pi_t + (\partial/\partial t)(\pi_t) = 0$$

If $\phi_t$ is the flow of $Y_t = -X$, defined by:

$$\frac{d}{dt}(\phi_t(x)) = X_t(\phi_t(x))$$

then we have: $(\phi_t)_* \pi_t = \pi_0$ [1]. Therefore $(\phi_1)_* \pi' = \pi. \square$

## 4 End of proofs

If in lemma 1, one Poisson structure comes from a symplectic structure, so does the other.

Indeed lemma 1 says that all the Poisson structures $\pi_t$ have the same rank, hence they are all inveritble, since $\pi_0$ is invertible. The family of symplectic forms joining $\Omega$ to $\Omega'$ is given by

$$\Omega_t(X, Y) = \pi_t((\tilde{\pi}_t)^{-1}(X), (\tilde{\pi}_t)^{-1}(Y)).$$

To complete the proof of theorem 1, we need to show that the symplectic forms have the same cohomology class. This is a consequence of the following

**Proposition 1** *If $G_t$ denotes the Poisson bivector coming from the symplectic forms $\Omega_t$, on a smooth manifold $M$ then the statements*

$$\phi_t^* \Omega_t = \Omega_0 \tag{4.1}$$

$$\phi_{t*} G_t = G_0. \tag{4.2}$$

*are equivalent. The last statement is equivalent to*

$$\dot{G}_t = -[X_t, G_t] \tag{4.3}$$

*where the family of vector fields $X_t$ integrates to produce the family of diffeomorphisms $\phi_t$, (in case $M$ is compact).*

**Proof** The second statement is equivalent to:

$$\frac{d}{dt}\phi_{t*}G_t = \phi_{t*}(L_{X_t}G_t + \dot{G}_t) = \phi_{t*}([X_t, G_t] + \dot{G}_t) = 0 .$$

We thus have :

$$\dot{G}_t = -[X_t, G_t] .$$

Let us first recall the definition of $\phi_* G$ for a diffeomorphism $\phi$: for $x \in M$, let $(\tau_\phi(x))$ be the transpose ( mapping $T_x^* M$ to $T_{\phi(x)}^* M$) of the differential $T_{\phi(x)}\phi^{-1}$ then:

$$(\phi_* G)(\alpha, \beta)(x) = G(\phi(x))(\tau_\phi(x).\alpha(x), \tau_\phi(x).\beta(x))$$

Let $G_t$ be the family of Poisson structures defined by the family of symplectic forms $\Omega_t$, i.e.:

$$\Omega_t(X, Y) = G_t(i(X)\Omega_t, i(Y)\Omega_t)$$

Assume that there exists a family of diffeomorphisms $\phi_t$ such that $\phi_t^*\Omega_t = \Omega_0$. We have:

$$((\phi_t)_*G_t)(\alpha, \beta)(x) = G_t(\phi_t(x))((\tau_{\phi_t}(x).\alpha(x), \tau_{\phi_t}(x).\beta(x))$$

$$= \Omega_t(\phi_t(x))(X_t, Y_t)$$

where $X_t$ is defined by the equality

$$\Omega_t(\phi_t(x))(X_t, \xi) = (\tau_{\phi_t}(x).\alpha(x))(\xi(x)) = (\phi_t^{-1})^*\alpha)(\xi)(\phi_t(x))$$

which means that

$$i(X_t)\Omega_t = (\phi_t^{-1})^*\alpha$$

and an analogous expression for $Y_t$. Therefore

$$\alpha = (\phi_t)^*[i(X_t)\Omega_t] = i((\phi_t^{-1})_*X_t)(\phi_t^*\Omega_t) = i((\phi_t^{-1})_*X_t)\Omega_0,$$

and an analogous expression for $\beta$. We have:

$$((\phi_t)_*G_t)(\alpha, \beta)(x) = \Omega_t(\phi_t(x))(X_t, Y_t) = ((\phi_t^{-1})^*\Omega_0)(\phi_t(x))(X_t, Y_t)$$

$$= \Omega_0(x)((\varphi_t^{-1})_*X_t, (\phi_t^{-1})_*Y_t)$$

$$= G_0(i((\phi_t^{-1})_*X_t)\Omega_0, i((\phi_t^{-1})_*X_t)\Omega_0) = G_0(\alpha, \beta)(x)$$

This shows that (4.1) implies (4.2). Reading this proof backward shows that (4.2) implies (4.1).$\square$

Theorem 2 is a consequence of the fact proved by Vaisman [7] saying that on a symmetric Riemannian manifold, the bivectors obtained as the musical images of harmonic 2-forms are compatible Poisson structures.

Observe that in theorem 2, the existence of $\phi$, isotopic to the identity, such that $\phi_*\pi = \pi'$ does not imply that the harmonic forms are exchanged by $\phi^*$. Otherwise, they would have the same cohomology class (homotopy invariance of the de Rham cohomology). Since they are harmonic, they must be equal (Hodge-de Rham theory).

## 5   Lifting of L-P coboundaries

On compact manifolds, Hodge-de Rham theory provides smooth liftings of de Rham coboundaries: i.e. if $\alpha_t$ is a smooth family of exact p-forms, there is a smooth family $\beta_t$ of (p-1)-forms such that $d\beta_t = \alpha_t$. This was the main argument in the proof of Moser's theorem cited above. Other methods of lifting de Rham coboundaries were given, among others, by Banyaga [2] and Laudenbach [3].

Since the spaces $\mathcal{V}^*(TM)$ are nuclear spaces [6], we may apply the arguments of [2] to the Poisson-Lichnerowicz cohomology provided that the space of coboundaries $\delta(\mathcal{V}^p(TM))$ is a closed subspace of the cycles $Z^{p+1} = Ker\delta$. Those arguments (page 186 of [2]) may be summarized, in this precise case, as follows:

Let us first recall some facts concerning tensor products of topological linear vector spaces (TVS). We refer to [6] for more details. There are two natural topologies on the tensor product $E \otimes F$ of two locally convex TVS. A space $E$ is said to be *nuclear* if, for any locally convex TVS $F$, both completions of $E \otimes F$ coincide; in this case the completion will be denoted by $E \widehat{\otimes} F$. For any open subset $U$ of $\mathbf{R}^n$, $C^\infty(U)$ is nuclear (with the topology of uniform convergence of all derivatives on compact subsets of $U$). Moreover, if $E$ is a complete locally convex TVS one has

$$C^\infty(U)\widehat{\otimes}E \simeq C^\infty(U,E) .$$

Let $\delta = [G_0,.]$ be the LP-coboundary operator. We denote by $B^2(\mathcal{V}^2(M)) = \delta(\mathcal{V}^1(M)) \subset \mathcal{V}^2(M)$ the subspace formed by the coboundaries. The space $\mathcal{V}^1(M))$ with its natural $C^\infty$-topology is a complete locally convex TVS. This is a standard result by the compactness of $M$. In case the second space $B^2(\mathcal{V}^2(M))$ is closed in $\mathcal{V}^2(M)$, one could reproduce, in the case of L-P-cohomology, the classical argument contained in [2]:

Let $I$ be an open neighborhood of zero in the real line. Consider now the surjection

$$\delta : \mathcal{V}^1(M) \longrightarrow B^2(\mathcal{V}^2(M))$$

which induces on the completed tensor product the following surjection

$$id\widehat{\otimes}\delta : C^\infty(I)\widehat{\otimes}\mathcal{V}^1(M) \longrightarrow C^\infty(I)\widehat{\otimes}B^2(\mathcal{V}^2(M)) .$$

Hence, the following arrow is still surjective

$$\delta : C^\infty(I,\mathcal{V}^1(M)) \longrightarrow C^\infty(I,B^2(\mathcal{V}^2(M))) .$$

The one–parameter smooth family $t \mapsto G_t - G_0$ is, by assumption, an element of $C^\infty(I, B^2(\mathcal{V}^2(M)))$. Therefore the above argument implies the existence of $(t \mapsto X_t) \in C^\infty(I, \mathcal{V}^1(M))$ such that $\delta(X_t) = [G_0, X_t] = G_t - G_0$. This will proves the following

**Theorem 3** *Let $(M, \pi)$ be a compact Poisson manifold such that the space of $\delta$-coboundaries is a closed subspace of the space of $\delta$-cocycles. Then given a smooth family of p-vectors $\lambda_t$ which are $\delta$-exact, there exists a smooth family of (p-1)- vectors $\mu_t$ such that $\delta\mu_t = \lambda_t$.*

The condition in theorem 3 is highly non trivial since the Lichnerowicz-Poisson cohomology is very complicated and may fail to be Hausdorff (when endowed with the quotient topology). However we can apply this theorem to deformations $\lambda_t$ of $\pi$, if $\pi$ is symplectic and get

**Theorem 4** *Let $(M, \Omega)$ be a compact symplectic manifold and $\pi$ the associated Poisson structure. Let $\lambda_t$ be a family of Poisson structures, with $\lambda_0 = \pi$, which are all in the same cohomology class of $H^2_\pi(M)$, then there exists a smooth family of vector fields $X_t$ such that*
$$(\partial/\partial t)\lambda_t = [\pi, X_t].$$

## References

1. R. Abraham, and J. Marsden, *Foundations of Mechanics*, The Benjamin/Cummings Publ Company, Reading, Massachussets (1978).
2. A. Banyaga, *Sur la structure du groupe des diffeomorphismes qui preservent une forme symplectique*, Comment. Math. Helv. 53(1978) 174-227.
3. F. Laudenbach , *Relevement lineaire des cobords*, Bul. Soc. Math. France, 111 (1983) 2, 147-150.
4. A. Lichnerowicz, *Les varietes de Poisson et leurs algebres de Lie associees*, J. Differential Geometry 12 (1977), 253-300.
5. J. Moser, *On the volume elements on a manifold*, Trans. Amer. Math. Soc. 120 (1965), 286-294.
6. F. Treves, *Topological vector spaces, distributions and kernels*, Academic Press, New York (1967)
7. I. Vaisman, *Lectures on the Geometry of Poisson Manifolds*, Birkhauser, Progress in Math. Vol 118 (1994)

# REMARKS ON ACTIONS ON COMPACTA BY SOME INFINITE-DIMENSIONAL GROUPS

VLADIMIR PESTOV

*School of Mathematical and Computing Sciences, Victoria University of Wellington, P.O. Box 600, Wellington, New Zealand* *
*E-mail: vova@mcs.vuw.ac.nz*
*http://www.mcs.vuw.ac.nz/~vova*

We discuss some techniques related to equivariant compactifications of uniform spaces and amenability of topological groups. In particular, we give a new proof of a recent result by Glasner and Weiss describing the universal minimal flow of the infinite symmetric group $\mathfrak{S}_\infty$ with the standard Polish topology, and extend Bekka's concept of an amenable representation, enabling one to deduce non-amenability of the Banach–Lie groups $GL(L_p)$ and $GL(\ell_p)$, $1 \leq p < \infty$.

## 1 Introduction

Let a topological group act continuously by uniform isomorphisms on a uniform space $X$. (One important situation is where $X = G/H$ is a homogeneous factor-space of $G$, equipped with the right uniform structure.) A compact space $K$, equipped with a continuous action of $G$, is called an *equivariant compactification* of $G$ if there is a uniformly continuous mapping $i \colon X \to K$ with dense image, commuting with the action of $G$. Compactifications of this type always exist, moreover every such $X$ admits a maximal $G$-equivariant compactification.

Here we discuss some ways in which equivariant compactifications can be used to study minimal actions and amenability of some infinite-dimensional groups. The latter term is used in an intuitive sense, to refer to concrete topological groups of importance in mathematics, such as, for instance, the full unitary groups of the infinite-dimensional Hilbert spaces. Some of these groups form infinite-dimensional Lie groups in one or other sense.

A topological group $G$ is called amenable if every compact $G$-space admits an invariant (regular Borel) probability measure. In particular, $G$ is *extremely amenable* if every compact $G$-space contains a fixed point (that is, admits an invariant Dirac measure). No non-trivial locally compact group is extremely amenable,[24] but among infinite-dimensional groups extreme amenability is not uncommon.[14,9,17,20,8,21]

---

*New permanent address beginning July 1, 2002: Department of Mathematics and Statistics, University of Ottawa, Ottawa, Ontario, K1N 6N5, Canada.

A continuous action of $G$ on a compact space $X$ is called minimal[1] if the $G$-orbit of every point $x \in X$ is everywhere dense in $X$. Every topological group $G$ possesses the universal minimal flow ($G$-space), $\mathcal{M}(G)$, such that every other minimal $G$-flow is a factor of $\mathcal{M}(G)$. For locally compact groups the size of the universal minimal flow is so immense that no constructive description is ever likely. (Cf. e.g. [5]) It comes as a surprise then that the universal minimal flow of at least some infinite-dimensional groups is manageable.

Moreover, it turns out that extremely amenable groups can be used as a tool in order to give an explicit description of the universal minimal flow $\mathcal{M}(G)$ even in cases where the flow is nontrivial. If a topological group $G$ contains a 'large' extremely amenable subgroup $H$, then the universal minimal flow of $G$ is a subflow of the equivariant compactification of the homogeneous space $G/H$, which is a much smaller object than $G$ itself. In some cases, it enables one to describe $\mathcal{M}(G)$. Such a technique was first used by the present author[17] in order to prove that the circle $\mathbb{S}^1$ forms the universal minimal flow for the group of orientation-preserving homeomorphisms of $\mathbb{S}^1$. Here we will use the argument in order to give a more transparent proof of the recent remarkable result by Glasner and Weiss,[10] who have characterized the universal minimal flow of the infinite symmetric group $\mathfrak{S}_\infty$, equipped with the standard Polish topology, as the compact space of all linear orders on $\mathbb{N}$. (The proof proposed here has an advantage that it extends the result beyond the separable case, to groups of permutations of an arbitrary infinite rank.)

Let us get back to the concept of an amenable topological group. A finer scale of 'shades of amenability' is given by the following concept: say that a homogeneous factor-space $G/H$ (or just a uniform $G$-space $X$) is amenable in the sense of Eymard[6] and Greenleaf[13] if the maximal equivariant compactification of $G/H$ supports an invariant probability measure.

Here is an important particular case. A unitary representation $\pi$ of a group $G$ in a Hilbert space $\mathcal{H}$ is amenable in the sense of Bekka[3] if there is a state on the von Neumann algebra of all bounded operators on $\mathcal{H}$, which is invariant under the action of $G$ by conjugations. It turns out[19] that a representation $\pi$ is amenable if and only if the unit sphere in the Hilbert space of the representation, upon which $G$ acts by isometries, is an amenable uniform $G$-space.

In general, it is more difficult to verify amenability of infinite-dimensional groups than that of locally compact or discrete ones, because some tools present in the locally compact case are missing. For example, if a locally compact group $G$ contains a closed copy of the free non-abelian group on two generators, then $G$ is non-amenable, because amenability is inherited by closed subgroups of locally compact groups. Not so beyond the locally compact case:[15] in fact, every topological group embeds into an extremely amenable group as a topological subgroup.[20] Another example: a locally compact group $G$ is amenable if and only if every strongly continuous unitary representation of $G$ is amenable.[3] For infinite-dimensional groups, neither implication need hold.

Here we show that in some situations the property of amenability is, in a sense, 'partly' inherited by topological subgroups.

We extend Bekka's concept as follows. Say that a representation $\pi$ of a group in a Banach space $E$ by bounded linear operators is amenable if the projective space of $E$ (upon which the group $G$ acts by isometries in a natural way) is an amenable $G$-space.

We show that every uniformly continuous representation of an amenable topological group is amenable. Since Eymard–Greenleaf amenability of an action (in particular, the Bekka amenability of a representation) of a group $G$ is clearly inherited by every subgroup $H < G$, we obtain a new possible way to prove that a topological group $G$ is non-amenable: to find a uniformly continuous representation $\pi$ of $G$ and a subgroup $H < G$ such that the restriction of $\pi$ to $H$ is apriori non-amenable.

The most natural class of infinite-dimensional groups admitting uniformly continuous representations are Banach–Lie groups and algebras of operators. As an illustration of our methods, we show that the general linear groups $GL(L_p)$ and $GL(\ell_p)$, where $1 \le p < \infty$, are non-amenable if equipped with the uniform operator topology. Even for Hilbert spaces this seems to be a new result, answering a question that Pierre de la Harpe asked me back in 1999.

## 2  Some abstract nonsense

### 2.1  Uniformities and compactifications

For a topological group $G$, we denote by $\mathcal{U}_r(G)$ the *Bourbaki-right* (= *Ellis-left*) uniform structure, whose entourage basis consists of the sets

$$V_r = \{(x,y) \in G \times G \mid xy^{-1} \in V\},$$

and $V$ runs over the neighbourhood filter, $\mathcal{N}_G$, of $G$ at the neutral element $e_G$. The symbol RUCB $(G)$ will denote the $C^*$-algebra of all Bourbaki right uniformly continuous bounded complex-valued functions on $G$, equipped with the supremum norm.

Denote by $\mathcal{S}(G)$ the Samuel compactification of the uniform space $(G, \mathcal{U}_r(G))$, that is, the maximal ideal space of RUCB $(G)$. This object (together with the distinguished point, $e = e_G$) is the well-known greatest ambit of $G$. ([1,4,18]) In other words, $\mathcal{S}(G)$ is a $G$-ambit (a compact $G$-space with a distinguished point having dense orbit), admitting a continuous equivariant map, preserving the distinguished points, to any other $G$-ambit.

Any two minimal compact $G$-subspaces of $\mathcal{S}(G)$ (whose existence is guaranteed by Zorn's lemma) are isomorphic as $G$-spaces.[1] (This is a non-trivial fact, because there is, in general, no *canonical* isomorphism.) This unique minimal $G$-space is denoted $\mathcal{M}(G)$ and called the *universal minimal $G$-space* (or *G-flow*).

Let $H$ be a (closed or not) subgroup of a topological group $G$. The *Bourbaki-right* uniform structure $\mathcal{U}_r(G/H)$ is by definition the finest uniform structure on $G/H$ making the factor-map

$$G \ni g \mapsto gH \in G/H$$

uniformly continuous if $G$ is equipped with the uniformity $\mathcal{U}_r(G)$. In general, the uniformity $\mathcal{U}_r(G/H)$ need not be separated even if $H$ is a closed subgroup, and the topology generated on $G/H$ by $\mathcal{U}_r(G/H)$ may be coarser than the factor-topology on $G/H$.

The standard action of $G$ on $G/H$ on the left extends to the action of $G$ on the Samuel compactification $\sigma(G/H, \mathcal{U}_r(G/H))$. (Notice that the Samuel compactification is always a separated uniform space, and so the compactification map need not be an embedding.) We will denote the latter compact space by $\mathcal{S}_H(G)$.

The Banach $G$-module $C(\mathcal{S}_H(G)) \cong \mathrm{UCB}\,(G/H, \mathcal{U}_r(G/H))$ embeds into the Banach $G$-module $\mathrm{UCB}\,(\mathcal{S}(G))$. Since the action of $G$ on the latter is well-known to be continuous, the same is true of the action of $G$ on the former Banach space (and $C^*$-algebra), and as a corollary, the action of $G$ on the compact space $\mathcal{S}_H(G)$ is continuous. With the image of the coset $H$ as the distinguished point, $\mathcal{S}_H(G)$ is thus a $G$-ambit.

## 2.2   Amenable groups and homogeneous spaces

A topological group $G$ is called *amenable* if one of the following equivalent conditions holds. All measures are assumed to be regular Borel.

1. There is a left-invariant mean on the space RUCB $(G)$.

2. There is an invariant probability measure on the greatest ambit $\mathcal{S}(G)$.

3. There is an invariant probability measure on every compact space upon which $G$ acts continuously.

4. There is an invariant probability measure on $\mathcal{M}(G)$.

See e.g. ([1], Chapter 12).

A topological group $G$ is called *extremely amenable* if one of the following equivalent conditions is true.

1. There is a multiplicative left-invariant mean on RUCB $(G)$.

2. There is a fixed point in $\mathcal{S}(G)$.

3. There is a fixed point in every compact space upon which $G$ acts continuously. (The fixed point on compacta property.)

4. The universal minimal flow $\mathcal{M}(G)$ is a singleton.

Even if this property looks exceedingly strong (in particular, no non-trivial locally compact group can possess it[24]), now we know numerous examples and entire classes of infinite-dimensional groups that are extremely amenable. The following list is not exhaustive: the unitary group of an infinite-dimensional Hilbert space with the strong operator topology;[14] the group of classes of measurable maps from the unit interval to the circle rotation group,[7,9] or, more generally, to any amenable locally compact group,[20] equipped with the topology of convergence in measure; the group of homeomorphisms of the closed (or open) unit interval with the compact-open topology;[17] the group of measure-preserving transformation of the standard Lebesgue measure space with the weak topology, as well as the group of measure class preserving transformations;[8] the group of isometries of the Urysohn universal metric space; [20] unitary groups of certain von Neumann algebras and $C^*$-algebras.[8]

If $H$ is a subgroup of a topological group $G$, then the homogeneous space $G/H$ (or the pair $(G, H)$) is *Eymard–Greenleaf amenable*[6,13] if there is a left-invariant mean on the space UCB $(G/H, \mathcal{U}_r(G/H))$. Equivalently, there exists an invariant probability measure on the ambit $\mathcal{S}_H(G)$.

More generally, one can talk of amenability of an action of a group $G$ on a uniform space $X$ by uniform isomorphisms. In such a situation, the topology on $G$ becomes irrelevant.

**Definition 2.1.** Let a group $G$ act by uniform isomorphisms on a uniform space $X$. Say that the action of $G$ is *Eymard–Greenleaf amenable*, or that $X$ is an *Eymard–Greenleaf amenable uniform $G$-space*, if there exists a $G$-invariant mean on the space UCB $(X)$. Equivalently (by the Riesz representation theorem), there exists an invariant probability measure on the Samuel compactification $\sigma X$.

For example, in the case $X = (G/H, \mathcal{U}_r(G/H))$ the above notion coincides with Eymard–Greenleaf amenability.

The following simple observation lends the concept some gravitas.

**Proposition 2.2.** *Every continuous action of an amenable locally compact group $G$ on a uniform space $X$ by uniform isomorphisms is amenable.*

*Proof.* Choose a point $x_0 \in X$ and set, for each $f \in \mathrm{UCB}(X)$ and every $g \in X$,

$$\tilde{f}(g) := f(gx_0).$$

The function $\tilde{f} \colon G \to \mathbb{C}$ so defined is bounded (obvious) and continuous, as the composition of two continuous maps: the orbit map $g \mapsto gx_0$ and the function $f \colon X \to \mathbb{C}$. Also, for each $h \in G$,

$$
\begin{aligned}
\widetilde{{}^{h}f}(g) &= {}^{h}f(gx_0) \\
&= f(h^{-1}gx_0) \\
&= \tilde{f}(h^{-1}g) \\
&= {}^{h}\tilde{f}(g),
\end{aligned}
$$

that is, the operator

$$\alpha \colon \mathrm{UCB}(X) \ni f \mapsto \tilde{f} \in \mathrm{CB}(G)$$

is $G$-equivariant. (Here $\mathrm{CB}(G)$ denotes the $C^*$-algebra of all bounded complex-valued continuous functions on $G$.) It is also clear that $\alpha$ is positive, linear, bounded of norm one, and sends the function $\mathbf{1}$ to $\mathbf{1}$. Since $G$ is amenable and locally compact, there exists a left-invariant mean $\phi$ on the space $\mathrm{CB}(G)$, and the composition $\phi \circ \alpha$ is a $G$-invariant mean on $\mathrm{UCB}(X)$. $\square$

This result is no longer true for more general topological groups, cf. a discussion in Subsection 4.3.

### 2.3 More on the ambit $\mathcal{S}_H(G)$

Let $G$ act continuously on a compact space $X$. Suppose there is a point $\xi \in X$ stabilized by $H$. The orbit map

$$G \ni g \mapsto g\xi \in X$$

then factors through the factor-space $G/H$, because for each $h \in H$ one has $(gh)\xi = g(h\xi) = g\xi$. Denote the resulting map $G/H \to X$ by $i$. Since the orbit map $G \to X$ is uniformly continuous relative to the uniformity $\mathcal{U}_r(G)$, the inductive definition of the uniformity $\mathcal{U}_r(G/H)$ implies that $i$ is uniformly continuous as well. Consequently, $i$ extends in a unique way to a continuous equivariant map $\mathcal{S}_H(G) \to X$. We conclude that $\mathcal{S}_H(G)$, with the distinguished point $H$ (the coset of $e_G$), is the universal compact $G$-ambit with the property that $H$ stabilizes the distinguished point.

In general, the compact $G$-space $\mathcal{S}_H(G)$ need not be minimal. The corresponding examples are easy to construct.

However, notice the following.

**Lemma 2.3.** *Let $G$ be a topological group, and let $H$ be a closed subgroup. Suppose the topological group $H$ is extremely amenable. Then any minimal compact $G$-subspace, $\mathcal{M}$, of $\mathcal{S}_H(G)$ is a universal minimal compact $G$-space.*

*Proof.* Let $X$ be an arbitrary minimal compact $G$-space. Because of extreme amenability of $H$, there is a point $\xi \in X$, stabilized by $H$. In view of the universality property of $\mathcal{S}_H(G)$ described above, there is a morphism of $G$-spaces $j \colon \mathcal{S}_H(G) \to X$ (taking $H$ to $\xi$). Because of minimality of $X$, the restriction of the map $j$ to $\mathcal{M}$ is onto $X$. We are done. $\square$

*Example* 2.4. Let $G = \mathrm{Homeo}_+(\mathbb{S}^1)$, the group of orientation-preserving homeomorphisms of the circle with the usual topology of uniform convergence, and let $H$ be the isotropy subgroup of any chosen element $\theta \in \mathbb{S}^1$. Then $H$ is isomorphic to the topological group $\mathrm{Homeo}_+[0,1]$ and therefore extermely amenable.[17] The right uniform factor-space $G/H$ is easily verified to be isomorphic to the circle $\mathbb{S}^1$ with the unique compatible uniformity, and therefore the ambit $\mathcal{S}_H(G)$ is $\mathbb{S}^1$ itself with the distinguished point $\theta$. Since it is obviously a minimal $G$-space, we conclude by Lemma 2.3: $\mathbb{S}^1$ is the universal minimal $\mathrm{Homeo}_+(\mathbb{S}^1)$-space. This fact, established by the present author in ([17]), was probably the first instance where a non-trivial universal minimal flow of any topological group has been computed explicitly.

Here is another consequence of Lemma 2.3, showing that the class of extremely amenable group is closed under extensions, similarly to the class of amenable groups. This result was established (through a direct proof) during author's discussion with Thierry Giordano and Pierre de la Harpe in April 1999, and is, thus, a joint result.

**Corollary 2.5.** *Let $H$ be a closed normal subgroup of a topological group $G$. If topological groups $H$ and $G/H$ are extremely amenable, then so is $G$.*

*Proof.* In this case, the ambit $\mathcal{S}_H(G)$ is just the greatest ambit $\mathcal{S}(G/H)$ of the factor-group, and it contains a fixed point since $G/H$ is extremely amenable. Now we conclude by Lemma 2.3. $\square$

# 3 The universal minimal flow of the infinite symmetric group

Here we use Lemma 2.3 in order to reprove a result by Glasner and Weiss[10] describing the universal minimal flow of the infinite symmetric group. An idea of this new proof was briefly sketched by us in ([21], Exercises 11 and 12), but appears here in any detail for the first time.

Let $X$ be an infinite set (countable or not), and let $G = \mathfrak{S}_X$ denote the full group of permutations of $X$, equipped with the topology of simple convergence on $X$ viewed as a discrete space. For countable $X$, this topology is well known to be Polish (separable completely metrizable).

Denote by $\mathrm{LO}_X$ the set of all linear orders on $X$, equipped with the (compact) topology induced from $\{0,1\}^{X \times X}$. (Here a linear order $\prec$ is identified with the characteristic function of the corresponding relation $\{(x,y) \in X \times X : x \prec y\}$.)

The group $\mathfrak{S}_X$ acts on $\mathrm{LO}_X$ by double permutations:

$$(x \,{}^{\sigma}\!\prec y) \Leftrightarrow (\sigma^{-1}x \prec \sigma^{-1}y)$$

for all $\prec \in \mathrm{LO}_X$, $\sigma \in \mathfrak{S}_X$, and $x, y \in X$. This action is continuous and minimal (an easy check).

A linear order $\prec$ on $X$ is called $\omega$-*homogeneous* if every finite subset $A \subset X$ can be mapped onto any other subset $B \subset X$ of the same cardinality by an order-preserving bijection (order automorphism) of $(X, \prec)$. In particular, it follows that $\prec$ is a dense order without least and greatest elements. (In the case where $X$ is countable, this condition is equivalent to $\omega$-homogeneity.)

Every infinite set $X$ supports an $\omega$-homogeneous linear order. (Here is one proof: $X$ can be given the structure of an ordered field, because it has the same cardinality as $\mathbb{Q}(X)$, the purely transcendental field extension of $\mathbb{Q}$, and the field $\mathbb{Q}(X)$ is well known to be linearly orderable. And every linearly ordered field is $\omega$-homogeneous due to the existence of piecewise-linear monotone maps.) Choose an arbitrary such order on $X$, say $\prec$.

Let $H = \mathrm{Aut}(\prec)$ be the subgroup of all permutations preserving the linear order $\prec$. The left factor-space $G/H = \mathfrak{S}_X/\mathrm{Aut}(\prec)$ can be identified with a certain collection of linear orders on $X$, namely those obtained from $\prec$ by a permutation. Denote this collection by $\mathrm{LO}_{\prec}$. Thus, $G/H \cong \mathrm{LO}_{\prec}$ embeds into $\mathrm{LO}_X$.

As every compact space, $\mathrm{LO}_X$ supports a unique compatible uniform structure. It induces a totally bounded uniform structure on $\mathrm{LO}_{\prec}$.

**Lemma 3.1.** *The uniform structure on $G/H \cong \mathrm{LO}_{\prec}$, induced from the compact space $\mathrm{LO}_X$, coincides with the right uniform structure $\mathcal{U}_r(\mathfrak{S}_X/\mathrm{Aut}(\prec))$.*

*Proof.* We want to show that the uniform structure on $G/H$, induced from the compact space $\mathrm{LO}_X$, is the finest uniform structure making the quotient map

$$\mathfrak{S}_X \to \mathfrak{S}_X/\mathrm{Aut}\,(\prec) \cong \mathrm{LO}_\prec$$

right uniformly continuous. The proof consists of two parts.

(1) The map $\sigma \mapsto {}^\sigma\!\prec$ is uniformly continuous.

Let $F = \{x_1, \ldots, x_n\} \subset \omega$ be any finite subset, determining the following standard basic entourage of the uniformity of $\mathrm{LO}_X$:

$$W_F := \{(<_1, <_2) \in \mathrm{LO}_X \times \mathrm{LO}_X : \; <_1 |_F = <_2 |_F\}.$$

Denote by $\mathrm{St}_F$ the common isotropy subgroup of all $x_i \in F$, that is,

$$\mathrm{St}_F := \{\sigma \in \mathfrak{S}_X \mid \sigma(x_i) = x_i, \;\; i = 1, 2, \ldots, n\}.$$

This $\mathrm{St}_F$ is an open subgroup of $\mathfrak{S}_X$ and in particular a (standard) open neighbourhood of the identity. As such, it determines an element of the Bourbaki-right uniformity $\mathcal{U}_r(\mathfrak{S}_X)$:

$$V_F := \{(\sigma, \tau) \in \mathfrak{S}_X \times \mathfrak{S}_X \mid \sigma\tau^{-1} \in \mathrm{St}_F\}.$$

In other words, $(\sigma, \tau) \in V_F$ iff for all $i = 1, 2, \ldots, n$ one has $\tau^{-1}x_i = \sigma^{-1}x_i$.

If now $(\sigma, \tau) \in V_F$, then for every $i, j = 1, 2, \ldots, n$ one has

$$x_i \; {}^\sigma\!\prec x_j \Leftrightarrow \sigma^{-1}x_i \prec \sigma^{-1}x_j$$
$$\Leftrightarrow \tau^{-1}x_i \prec \tau^{-1}x_j$$
$$\Leftrightarrow x_i \; {}^\tau\!\prec x_j,$$

meaning that the restrictions of the orders ${}^\sigma\!\prec$ and ${}^\tau\!\prec$ to $F$ coincide, and thus $({}^\sigma\!\prec, {}^\tau\!\prec) \in W_F$.

(2) The image of the entourage $V_F$ under the (Cartesian square of) the map $\sigma \mapsto {}^\sigma\!\prec$ is *exactly* all of $W_F \cap (\mathrm{LO}_\prec \times \mathrm{LO}_\prec)$.

Indeed, suppose $(<_1, <_2) \in W_F \cap (\mathrm{LO}_\prec \times \mathrm{LO}_\prec)$, that is, $<_1$ and $<_2$ are linear orders on $X$, obtained from $\prec$ by suitable permutations, and whose restrictions to a finite subset $F$ coincide.

Choose two permutations $\sigma, \tau \in \mathfrak{S}_X$ such that $<_1 = {}^\sigma\!\prec$ and $<_2 = {}^\tau\!\prec$. For each $i = 1, 2, \ldots, n$, one necessarily has

$$\sigma^{-1}x_i = \tau^{-1}x_i$$

(if it were not so, then the orders ${}^\sigma\!\prec$ and ${}^\tau\!\prec$ would differ on $F$). Consequently, $(\sigma, \tau) \in V_F$.

Now we conclude that every uniform structure, $\mathcal{U}$, on $\mathrm{LO}_\prec$ that makes the map

$$(\mathfrak{S}_X, \mathcal{U}_r) \ni \sigma \mapsto {}^\sigma \prec \in (\mathrm{LO}_\prec, \mathcal{U})$$

uniformly continuous, must be coarser than the restriction of the uniformity of $\mathrm{LO}_X$ to $\mathrm{LO}_\prec$. Indeed, for every element $W \in \mathcal{U}$ there is, by the assumed uniform continuity of the above map, a finite $F \subset \omega$ with the image of $V_F$ contained in $W$, that is, with $W_F \subseteq W$. This accomplishes the argument. $\square$

**Lemma 3.2.** *The ambit $S_{\mathrm{Aut}\,(\prec)}(\mathfrak{S}_X)$ is isomorphic to $\mathrm{LO}_X$, with the distinguished element $\prec$.*

*Proof.* By Lemma 3.1, $(\mathfrak{S}_X/\mathrm{Aut}\,(\prec), \mathcal{U}_r(\mathfrak{S}_X/\mathrm{Aut}\,(\prec)))$ embeds into $\mathrm{LO}_X$ as a uniform subspace and an $\mathfrak{S}_X$-subspace. Also, $\mathrm{LO}_\prec$ is everywhere dense in $\mathrm{LO}_X$. As a consequence, the Samuel compactification of the precompact uniform space $(\mathfrak{S}_X/\mathrm{Aut}\,(\prec), \mathcal{U}_r(\mathfrak{S}_X/\mathrm{Aut}\,(\prec)))$ is simply its completion, that is, $\mathrm{LO}_X$. $\square$

An application of Lemma 2.3 (bearing in mind that the topological group $H = \mathrm{Aut}\,(\prec)$ is extremely amenable [17]) yields immediately:

**Theorem 3.3 (Glasner and Weiss [10]).** *The compact space $\mathrm{LO}_X$ forms the universal minimal $\mathfrak{S}_X$-space.* $\square$

*Remark 3.4.* The original theorem by Glasner and Weiss was established in the case of countable $X$. Our proof remains true for symmetric groups of arbitrary infinite rank.

*Remark 3.5.* The group $\mathfrak{S}_X$ contains, as a dense subgroup, the union of the directed family of permutation subgroups of finite rank, and consequently it is amenable. As a result, there is an invariant probability measure on the compact set $\mathrm{LO}_X$. Glasner and Weiss[10] have proved that such a measure is unique, that is, the action by $\mathfrak{S}_X$ on $\mathcal{M}(\mathfrak{S}_X) \cong \mathrm{LO}_X$ is uniquely ergodic.

Their argument can be made quite elementary (no Ergodic Theorem!) as follows. Let $\mu$ be a $\mathfrak{S}_X$-invariant probability measure on $\mathrm{LO}_X$. If $F \subset X$ is a finite subset, then every linear order, $<$, on $F$ determines a cylindrical subset

$$C_< := \{\prec \in \mathrm{LO}_X : \prec |_F = <\} \subset \mathrm{LO}_X.$$

Every two sets of this form, corresponding to different orders on $F$, are disjoint and can be taken to each other by a suitable permutation. As there are $n!$ of such sets, where $n = |F|$, the $\mu$-measure of each of them must equal $1/n!$. Consequently, the functional $\int d\mu$ is uniquely defined on the characteristic functions of cylinder sets $C_<$, which functions are continuous and separate points, because sets $C_<$ are open and closed and form a basis of open subsets of $LO_X$. Now the Stone–Weierstrass theorem implies uniqueness of $\int d\mu$ on all of $C(LO_X)$.

*Remark 3.6.* Every extremely amenable subgroup $H$ of $\mathfrak{S}_X$ is contained in one of the subgroups of the form Aut $(\prec)$. (Indeed, $H$ must possess a fixed point in the space $LO_X$, that is, preserve a linear order $\prec$ on $X$.)

At the same time, not every subgroup of the form Aut $(\prec)$ is extremely amenable. For example, if the linear order $\prec$ is such that for some cover of $X$ by three disjoint convex subsets $A, B, C$ one has $A < B < C$, $A$ and $C$ are densely ordered, and $B$ has type $\mathbb{Z}$, then the group Aut $(\prec)$ is topologically isomorphic to the product of three groups of order automorphisms, and since Aut $(B) \cong \mathbb{Z}$ is not extremely amenable, neither is Aut $(\prec)$.

On the other hand, a similar construction can be used to produce examples of groups of type Aut $(\prec)$ which are extremely amenable even if the order $\prec$ is not dense (admits gaps).

*Example 3.7.* The *tame topology* on the group $U(\infty) = \cup_{i=1}^{\infty} U(n)$ is the topology of simple convergence on the sphere $\mathbb{S}(\infty) = \cup_{i=1}^{\infty} \mathbb{S}^n$ (the intersection of the unit sphere of $\ell_2$ with the direct limit space $\mathbb{C}^\infty$), viewed as discrete. Thus, $U(\infty)$ receives the subgroup topology from $\mathfrak{S}_{\mathbb{S}(\infty)}$. This topology is of considerable interest in representation theory of the infinite unitary group,[16] where unitary representations strongly continuous with regard to the tame topology are called *tame representations*.

As a consequence of the Remark 3.6, the group $U(\infty)$ with the tame topology is not extremely amenable: indeed, it is easy to see that no linear order on $\mathbb{S}(\infty)$ is preserved by all operators from $U(\infty)$. Thus, the universal minimal flow $\mathcal{M}(U(\infty)_{tame})$ is nontrivial.

*Remark 3.8.* Let a group $G$ act by uniform isomorphisms on a uniform space $X$. The pair $(G, X)$ has the *Ramsey–Dvoretzky–Milman property* if for every bounded uniformly continuous function $f$ from $X$ to a finite-dimensional Euclidean space, every finite $F \subseteq X$, and each $\varepsilon > 0$ there is a $g \in G$ such that the oscillation of $f$ on the translate $gF$ is less than $\varepsilon$. This concept links extreme amenability with Ramsey theory, because a topological group $G$ is extremely amenable if and only if every continuous transitive action of $G$ by isometries on a metric space has the Ramsey–Dvoretzky–Milman property.[21]

The statement in Example 3.7 can be strengthened: a result by Graham[11] on the so-called sphere-Ramsey spaces implies that already the pair $(U(\infty), \mathbb{S}(\infty))$, where $\mathbb{S}(\infty)$ is equipped with the discrete ($\{0, 1\}$-valued) metric, does not have the Ramsey–Dvoretzky–Milman property.

This sort of dynamical properties, formulated for appropriate groups of affine transformations, is linked to the central open question of Euclidean Ramsey theory: is every finite spherical metric space Ramsey? [12]

## 4 Amenable representations

### 4.1 The projective space

Let $E$ be a (complex or real) Banach space. Denote by $\mathbb{P}_E$ the projective space of $E$. If we think of $\mathbb{P}_E$ as a factor-space of the unit sphere $\mathbb{S}_E$ of $E$, then $\mathbb{P}_E$ becomes a metric space via the rule

$$d(x,y) = \inf\{\|\xi - \zeta\| : \xi, \zeta \in \mathbb{S}_{\mathcal{H}}, p(\xi) = x, p(\zeta) = y\},$$

where $p: \mathbb{S}_E \to \mathbb{P}_E$ is the canonical factor-map. Notice that the infimum in the formula above is in fact minimum. The proof of the triangle inequality is based on the invariance of the norm distance on the sphere under multiplication by scalars. The above metric on the projective space is complete.

Let $T \in \mathrm{GL}\,(E)$ be a bounded linear invertible operator on a Banach space $E$. Define a mapping $\tilde{T}$ from the projective space $\mathbb{P}_E$ to itself as follows: for every $\xi \in \mathbb{S}_E$ set

$$\tilde{T}(p(\xi)) = p\left(\frac{T(\xi)}{\|T(\xi)\|}\right).$$

The above definition is clearly independent on the choice of a representative, $\xi$, of an element of the projective space $x \in \mathbb{P}_E$.

**Lemma 4.1.** *The mapping $\tilde{T}$ is a uniform isomorphism (and even a bi-Lipschitz isomorphism) of the projective space $\mathbb{P}_E$.*

*Proof.* It is enough to show that $\tilde{T}$ is uniformly continuous, because $\widetilde{TS} = \tilde{T}\tilde{S}$ and so $\widetilde{T^{-1}} = \tilde{T}^{-1}$. Let $x, y \in \mathbb{P}_E$, and let $\xi, \zeta \in \mathbb{S}_E$ be such that $p(\xi) = x$, $p(\zeta) = y$, and $\|\xi - \zeta\| = d(x, y)$. Both $\|T(\xi)\|$ and $\|T(\zeta)\|$ are bounded below by $\|T^{-1}\|^{-1}$, and therefore

$$d(\tilde{T}(x), \tilde{T}(y)) \le \frac{\pi}{2} \|T^{-1}\| \cdot \|T(\xi) - T(\zeta)\|$$

$$\le \frac{\pi}{2} \|T^{-1}\| \cdot \|T\| \cdot \|\xi - \zeta\|$$

$$= \frac{\pi}{2} \|T^{-1}\| \cdot \|T\| \, d(x, y).$$

$\square$

Let us recall the following notion from theory of transformation groups.

**Definition 4.2.** Let a group $G$ act by uniform isomorphisms on a uniform space $X = (X, \mathcal{U}_X)$. The action is called *bounded* (or else *motion equicontinuous*) if for every $U \in \mathcal{U}_X$ there is a neighbourhood of the identity, $V \ni e_G$, such that $(x, g \cdot x) \in U$ for all $g \in V$ and $x \in X$.

Notice that every bounded action is continuous.

*Example* 4.3. The action of $\mathrm{GL}(E)$ on the unit sphere $\mathbb{S}_E$ (and moreover the unit ball) of a Banach space $E$ is bounded, by the very definition of the uniform operator topology.

**Lemma 4.4.** *The correspondence*

$$\mathrm{GL}(E) \ni T \mapsto \tilde{T}$$

*determines an action of the general linear group* $\mathrm{GL}(E)$ *on the projective space* $\mathbb{P}_E$ *by uniform isomorphisms. With respect to the uniform operator topology on* $\mathrm{GL}(E)$, *the action is bounded.*

*Proof.* The first part of the statement is easy to check using Lemma 4.1. As to the second, if $\|T - \mathbb{I}\| < \varepsilon$, then for every $\xi \in \mathbb{S}_E$

$$\left\| \tilde{T}(x) - x \right\| \leq \frac{\pi}{2} \left\| T^{-1} \right\| \cdot \|T(x) - x\|$$
$$< \frac{\pi \varepsilon}{2(1 - \varepsilon)}.$$

$\square$

### 4.2  Extension of Bekka's amenability

We want to reformulate the concept of an amenable representation in the sense of Bekka in order to present a natural extension of it.

Let $\pi$ be a unitary representation of a group $G$ in a Hilbert space $\mathcal{H}$. One says that $\pi$ is *amenable* [3] if there exists a state, $\phi$, on the von Neumann algebra $\mathcal{B}(\mathcal{H})$ of all bounded operators on the space $\mathcal{H}$ of representation, which is invariant under the action of $G$ by inner automorphisms: $\phi(\pi(g)T\pi(g)^{-1}) = \phi(T)$ for every $T \in B(\mathcal{H})$ and every $g \in G$.

The group $G$ acts on the unit sphere $\mathbb{S}_\mathcal{H}$ by isometries, and it was shown by the author [19] that a unitary representation $\pi$ of a group $G$ in a Hilbert space $\mathcal{H}$ is amenable if and only if $\mathbb{S}_\mathcal{H}$ is an amenable uniform $G$-space in the sense of our Definition 2.1, that is, there exists a $G$-invariant mean on the space $\mathrm{UCB}\,(\mathbb{S}_\mathcal{H})$ or, equivalently, an invariant probability measure on the Samuel compactification of the sphere $\mathbb{S}_\mathcal{H}$.

While the implication $\Rightarrow$ is based on some results obtained by Bekka using deep techniques by Connes, the implication $\Leftarrow$ is elementary. We need to reproduce it here.

$\triangleleft$ Let $\psi$ be a $G$-invariant mean on UCB $(\mathbb{S}_{\mathcal{H}})$. Every bounded linear operator $T$ on $\mathcal{H}$ defines a bounded uniformly continuous (in fact, even Lipschitz) function $f_T \colon \mathbb{S}_{\mathcal{H}} \to \mathbb{C}$ by the rule

$$\mathbb{S}_{\mathcal{H}} \ni \xi \mapsto f_T(\xi) := \langle T\xi, \xi \rangle \in \mathbb{C}.$$

Now set $\phi(T) := \psi(f_T)$. This $\phi$ is a $G$-invariant state on $\mathcal{B}(\mathcal{H})$. $\triangleright$

Notice that the function $f_T$ in the proof above is symmetric: for every $\lambda \in \mathbb{C}$, $|\lambda| = 1$, and each $\xi \in \mathbb{S}_\pi$, one has $f_T(\lambda\xi) = f_T(\xi)$. In other words, $f_T$ is constant on the preimages of $p$. Consequently, $f_T$ factors through a function $\tilde{f}_T$ on the projective space $\mathbb{P}_{\mathcal{H}}$; clearly, $\tilde{f}_T$ is also uniformly continuous and bounded. It means that the above proof only uses the existence of a $G$-invariant mean on the function space UCB $(\mathbb{P}_{\mathcal{H}})$.

On the other hand, the Banach space (and $G$-module) UCB $(\mathbb{P}_{\mathcal{H}})$ admits an obvious equivariant embedding into UCB $(\mathbb{S}_{\mathcal{H}})$; namely, it can be identified with the Banach $G$-submodule of all functions symmetric in the above sense. The restriction of a $G$-invariant mean from UCB $(\mathbb{S}_{\mathcal{H}})$ to UCB $(\mathbb{P}_{\mathcal{H}})$ is again a $G$-invariant mean.

We have thus established the following.

**Theorem 4.5.** *A unitary representation $\pi$ of a group $G$ in a Hilbert space $\mathcal{H}$ is amenable if and only if the projective space $\mathbb{P}_{\mathcal{H}}$ is an Eymard–Greenleaf amenable uniform $G$-space.* $\square$

The advantage of this reformulation is that it allows for an extension of the concept of an amenable representation to group representations by bounded linear operators that are not necessarily unitary.

**Definition 4.6.** Say that a representation $\pi$ of a group $G$ by bounded linear operators in a normed space $E$ is *amenable* if the action of $G$ by uniform isometries on the projective space $\mathbb{P}_E$, associated to $\pi$ as in Lemma 4.4, is Eymard–Greenleaf amenable in the sense of Definition 2.1.

**Theorem 4.7.** *Let $G$ be a locally compact group, and let $\mu$ be a quasi-invariant measure on $G$. Let $1 \le p < \infty$. The left quasi-regular representation of $G$ in $L_p(\mu)$ is amenable if and only if $G$ is amenable.*

*Proof.* The left quasi-regular representation, $\gamma$, of $G$ in $L_p(\mu)$ (cf. e.g. [23]) is given by the formula

$$^g f(x) = \left( \frac{d(\tau \circ g^{-1})}{d\tau} \right)^{\frac{1}{p}} f(g^{-1}x),$$

where $d/d\tau$ is the Radon-Nykodim derivative. It is a strongly continuous representation by isometries. Necessity ($\Rightarrow$) thus follows at once from Proposition 2.2.

To prove sufficiency ($\Leftarrow$), assume $\gamma$ is amenable. Then there exists an invariant mean, $\phi$, on UCB $(\mathbb{S}_p)$, where $\mathbb{S}_p$ stands for the unit sphere in $L_p(\mu)$. For every Borel subset $A \subseteq G$, define a function $f_A \colon \mathbb{S}_p \to \mathbb{C}$ by letting for each $\xi \in \mathbb{S}_p$

$$f_A(\xi) = \|\xi \cdot \chi_A\|^p,$$

where $\chi_A$ is the characteristic function of $A$. The function $f_A$ is bounded and uniformly continuous on $\mathbb{S}_p$. For every $g \in G$,

$$
\begin{aligned}
{}^g f_A(\xi) &= f_A\left({}^{g^{-1}}\xi\right) \\
&= \int_A \left| {}^{g^{-1}}\xi(x) \right|^p d\mu(x) \\
&= \int_A \frac{d\mu \circ g}{d\mu} \, |\xi(gx)|^p \, d\mu(x) \\
&= \int_{gA} |\xi(y)|^p \, d\mu(y) \\
&= f_{gA}(\xi),
\end{aligned}
$$

that is, ${}^g f_A = f_{gA}$. It is now easily seen that $m(A) := \phi(f_A)$ is a finitely additive left-invariant measure on $G$, vanishing on locally null sets, and so $G$ is amenable. $\square$

## 4.3 Uniformly continuous representations

In contrast with Proposition 2.2, even a strongly continuous unitary representation of an amenable non-locally compact topological group need not be amenable. The simplest example is the standard representation of the full unitary group $U(\mathcal{H})_s$ of an infinite-dimensional Hilbert space, equipped with the strong topology. It is not amenable because it contains, as a subrepresentation, the left regular representation of a free nonabelian group, which is not amenable. (The left regular representation of a locally compact group $G$ is amenable if and only if $G$ is amenable,[3] cf. also Theorem 4.7.)

A part of the story here is that when a topological group $G$ acts continuously by uniform isomorphisms on a uniform space $X$, the resulting representation of $G$ by isometries in UCB $(X)$ need not be continuous. (This is the case, for instance, in the same example $G = U(\mathcal{H})_s$, $X = \mathbb{S}_{\mathcal{H}}$.) Equivalently, the extension of the action of $G$ to the Samuel (uniform) compactification $\sigma X$ is discontinuous, and therefore one cannot deduce the existence of an invariant measure on $\sigma X$ from the assumed amenability of $G$.

Here is a simple case where the continuity of action is assured.

**Lemma 4.8.** *Suppose a topological group $G$ acts in a bounded way on a uniform space $X$. Then the resulting representation of $G$ in* UCB $(X)$, *as well as the action of $G$ on $\sigma X$, are both continuous.*

*Proof.* Since $G$ acts on UCB $(X)$ by isometries, it is enough to show that the mapping $G \ni g \mapsto {}^g f \in$ UCB $(X)$ is continuous at identity. By a given $\varepsilon > 0$, choose a $U \in \mathcal{U}_X$ using the uniform continuity of $f$, and a symmetric neighbourhood $V \ni e_G$ so that $|f(x) - f(y)| < \varepsilon$ whenever $(x, y) \in U$ and $(x, g \cdot x) \in U$ once $g \in V$ and $x \in X$. Now for each $x \in X$, $|f(x) - f(g^{-1}x)| \le \varepsilon$ once $g \in V$, that is, $\|f - {}^g f\|_{sup} < \varepsilon$, and we are done.

Now recall that the Samuel (maximal uniform) compactification of $X$ is the maximal ideal space of UCB $(X)$. It is a simple observation (which was made, for instance, by Teleman[22]) that a representation of a topological group by isomorphisms of a commutative $C^*$-algebra is strongly continuous if and only if the associated action of $G$ on the maximal ideal space is continuous. $\square$

The following three corollaries are immediate.

**Corollary 4.9.** *Let a topological group $G$ act in a bounded way on a uniform space $X$. If $G$ is amenable, then $X$ is an Eymard–Greenleaf amenable uniform $G$-space.* $\square$

**Corollary 4.10.** *Let $\pi$ be a uniformly continuous representation of a topological group in a Banach space $E$. If $G$ is amenable, then $\pi$ is an amenable representation.* $\square$

**Corollary 4.11.** *Let $E$ be a Banach space, and let $G$ be a topological subgroup of* GL $(E)$ *(equipped with the uniform operator topology). If $H$ is a subgroup of $G$ and the restriction of the standard representation of* GL $(E)$ *in $E$ to $H$ is non-amenable, then $G$ is a non-amenable topological group.* $\square$

*4.4 Groups of operators*

**Theorem 4.12.** *The general linear groups* GL $(L_p)$ *and* GL $(\ell_p)$, $1 \le p < \infty$, *with the uniform operator topology are non-amenable.*

*Proof.* Both spaces $L_p$ and $\ell_p$ can be realized as $L_p(\mu)$, where $\mu$ is a quasi-invariant measure on a non-amenable locally compact group, $H$. (For instance, $H = $ SL $(2, \mathbb{C})$ for the continuous case and $H = $ SL $(2, \mathbb{Z})$ for the purely atomic one.) Identify $H$ with an (abstract, non-topological) subgroup of GL $(L_p(\mu))$ via the left quasi-regular representation, $\gamma$. The restriction of the standard representation of GL $(L_p(\mu))$ to $H$ is $\gamma$, which is a non-amenable representation (Theorem 4.7), and Corollary 4.11 applies. □

It is interesting to compare the above result with the following.

**Theorem 4.13.** *The isometry group* Iso($\ell_p$), $1 \leq p < \infty$, $p \neq 2$, *equipped with the strong operator topology, is amenable, but not extremely amenable.*

*Proof.* The isometry groups in question, as abstract groups, have been described by Banach in his classical 1932 treatise [2] (Chap. XI, §5, pp. 178–179). For $p > 1$, $p \neq 2$ the group Iso($\ell_p$) is isomorphic to the semidirect product of the group of permutations $\mathfrak{S}_\infty$ and the countable power $U(1)^{\mathbb{N}}$ (in the complex case) or $\{1,-1\}^{\mathbb{Z}}$ (in the real case). Here the group of permutations acts on $\ell_p$ by permuting coordinates, while the group of sequences of scalars of absolute value one acts by coordinate-wise multiplication. The semidirect product is formed with regard to an obvious action of $\mathfrak{S}_\infty$ on $U(1)^{\mathbb{N}}$ (in the real case, $\{1,-1\}^{\mathbb{N}}$).

The strong operator topology restricted to the group $\mathfrak{S}_\infty$ is the standard Polish topology, and restricted to the product group, it is the standard product topology. Thus, Iso($\ell_p$) $\cong \mathfrak{S}_\infty \ltimes U(1)^{\mathbb{N}}$ (correspondingly, $\mathfrak{S}_\infty \ltimes \{1,-1\}^{\mathbb{N}}$) is the semidirect product of a Polish group with a compact metric group. Since $\mathfrak{S}_\infty$ is an amenable topological group, so is Iso($\ell_p$). Since the non-extremely amenable group $\mathfrak{S}_\infty$ is a topological factor-group of Iso($\ell_p$), the latter group is not extremely amenable either. □

*Remark* 4.14. Using the description of the universal minimal flow of $\mathfrak{S}_\infty$ due to Glasner and Weiss discussed in Section 3, as well as some standard means of uniform topology, one can show that the universal minimal flow of the topological group Iso($\ell_p$) is homeomorphic to the product of the compact space LO$_\infty$ of all linear orders on the natural numbers with the compact group $U(1)^{\mathbb{N}}$ (complex case) or $\{1,-1\}^{\mathbb{N}}$ (real case). This compact space is equipped with a skew product action:

$$(\sigma, f) \cdot (\prec, g) = (^\sigma\prec, f \cdot {}^\sigma g),$$

where $^\sigma g(n) = g(\sigma^{-1}n)$. Again, this action is uniquely ergodic. We leave the details out.

*Remark* 4.15. By contrast, for $p = 2$ the unitary group of an infinite-dimensional Hilbert space with the strong operator topology is extremely amenable. This is due to Gromov and Milman[14].

We conjecture that the groups $\mathrm{Iso}(L_p)$, $1 \leq p < \infty$, with the strong operator topology are all extremely amenable.

## Acknowledgments

I am grateful to Joshua Leslie and Thierry Robart for their hospitality during the conference on Infinite Dimensional Lie Groups in Geometry and Representation Theory and for their patience during the preparation of this volume. Stimulating conversations with Thierry Giordano, Eli Glasner, Pierre de la Harpe, and Michael Megrelishvili are acknowledged. The present research has been supported by the Marsden Fund of the Royal Society of New Zealand through the grant project 'Geometry of high-dimensional structures: dynamical aspects.'

## References

1. J. Auslander, *Minimal Flows and Their Extensions*, North-Holland Mathematics Studies **153**, North-Holland, Amsterdam–NY–London–Tokyo, 1988.
2. S. Banach, *Théorie des opérations linéaires*, Second edition, Chelsea Publishing Co., New York. (Reprint of the first edition, 1932, with corrections and an addendum.)
3. M.E.B. Bekka, *Amenable unitary representations of locally compact groups*, Invent. Math. **100** (1990), 383–401.
4. R.B. Brook, *A construction of the greatest ambit*, Math. Systems Theory **4** (1970), 243–248.
5. R. Ellis, *Universal minimal sets*, Proc. Amer. Math. Soc. **11** (1960), 540–543.
6. P. Eymard, *Moyennes invariantes et représentations unitaires*, Lecture Notes Math. **300**, Springer-Verlag, Berlin-New York, 1972.
7. H. Furstenberg and B. Weiss, unpublished.
8. T. Giordano and V. Pestov, *Some extremely amenable groups*, C.R. Acad. Sci. Paris, Sér. I, to appear (2002).
   arXiv e-print: http://arXiv.org/abs/math.GR/0109138
9. S. Glasner, *On minimal actions of Polish groups*, Top. Appl. **85** (1998), 119–125.
10. E. Glasner and B. Weiss, *Minimal actions of the group $\mathbb{S}(\mathbb{Z})$ of permutations of the integers*, Geom. Funct. Anal., to appear (2002).
11. R.L. Graham, *Euclidean Ramsey theorems on the n-sphere*, J. Graph Theory **7** (1983), 105–114.

12. R.L. Graham, *Recent trends in Euclidean Ramsey theory*, Discrete Math. **136** (1994), 119–127.

13. F.P. Greenleaf, *Amenable actions of locally compact groups*, J. Funct. Anal. **4** (1969), 295–315.

14. M. Gromov and V.D. Milman, *A topological application of the isoperimetric inequality*, Amer. J. Math. **105** (1983), 843–854.

15. P. de la Harpe, *Moyennabilité de quelques groupes topologiques de dimension infinie*, C.R. Acad. Sci. Paris, Sér. A **277** (1973), 1037–1040.

16. G.I. Ol'shanskij, *Representations of infinite-dimensional classical groups, limits of enveloping algebras, and Yangians*, Topics in representation theory, Adv. Sov. Math. **2** (1991), 1-66.

17. V.G. Pestov, *On free actions, minimal flows, and a problem by Ellis*, Trans. Amer. Math. Soc. **350**, 4149–4165 (1998).

18. V. Pestov, *Some universal constructions in abstract topological dynamics*, in: Topological Dynamics and its Applications. A Volume in Honor of Robert Ellis, Contemp. Math. **215** (1998), 83–99.

19. V.G. Pestov, *Amenable representations and dynamics of the unit sphere in an infinite-dimensional Hilbert space.* – Geometric and Functional Analysis **10** (2000), 1171–1201.

20. V. Pestov, *Ramsey–Milman phenomenon, Urysohn metric spaces, and extremely amenable groups*, Israel Journal of Mathematics, to appear (2002).
arXiv e-print: http://arXiv.org/abs/math.FA/0004010

21. V. Pestov, *mm-Spaces and group actions*, L'Enseignement Mathématique, to appear (2002).
arXiv e-print: http://arXiv.org/abs/math.FA/0110287

22. S. Teleman, *Sur la représentation linéaire des groupes topologiques*, Ann. Sci. Ecole Norm. Sup. **74** (1957), 319–339.

23. G. Warner, *Harmonic Analysis on Semi-Simple Lie Groups I*, Grundlehren der math. Wissenschaften Bd. **188**, Springer-Verlag, Berlin–Heidelberg–NY, 1972.

24. W.A. Veech, *Topological dynamics*, Bull. Amer. Math. Soc. **83** (1977), 775–830.

www.ingramcontent.com/pod-product-compliance
Lightning Source LLC
Chambersburg PA
CBHW050642190326
41458CB00008B/2379

9 789812 380685